BLAST OFF

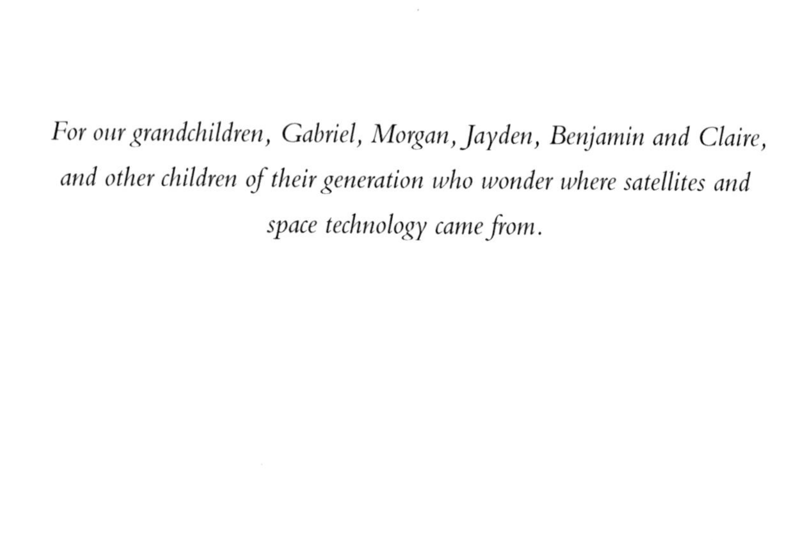

For our grandchildren, Gabriel, Morgan, Jayden, Benjamin and Claire, and other children of their generation who wonder where satellites and space technology came from.

Ken McCracken

BLAST OFF

Scientific adventures at the dawn of the Space Age

This edition published in the USA in 2008
New Holland Publishers (Australia) Pty Ltd
Sydney • Auckland • London • Cape Town
www.newholland.com.au

1/66 Gibbes Street Chatswood NSW 2067 Australia
218 Lake Road Northcote Auckland New Zealand
86 Edgware Road London W2 2EA United Kingdom
80 McKenzie Street Cape Town 8001 South Africa

National Library of Australia Cataloguing-in-Publication Data:

McCracken, K. G. (Kenneth G.), 1933-
Blast off : the dawn of the space age / author, Ken McCracken.

ISBN: 78174110 6442 (pbk.)

United States. National Aeronautics and Space Administration
History. Space race. Space sciences. Scientists.

500.5

Publisher: Martin Ford
Managing Editor: Lliane Clarke
Production Manager: Linda Bottari
Project Editor: Michael McGrath
Editor: Kirsten Chapman
Designer: Hayley Norman
Printer: Publisher's Graphics, USA

Cover image: The launch of Pioneer 7 *from Cape Canaveral, Florida, on 17 August, 1966. Insert. The* Pioneer 6 *(through 9) spacecraft. The experiments look out at space through the 'belly band', while the magnetic sensors are at the end of one of the long arms (both images courtesy NASA).*

Acknowledgements

Images as identified in the captions are used with the permission of the United States National Aeronautics and Space Administration (NASA) and the Defence Science and Technology Organization of Australia. Figures 3, 5, and 7 are reproduced by permission of the American Geophysical Union. Figures 4 and 6 are reproduced by kind permission of Springer Science and Business Media. Figure 1 is reproduced with the permission of the American Physical Society. Figure 2 is published by kind permission of M A Shea and D F Smart.

'The Meson Song' was recited and distributed to the attendees at the Echo Lake Conference on Cosmic Rays in 1949, and reproduced, *inter alia,* in *Early History of Cosmic Ray Studies,* Sekido and Elliot, Astrophysics and Space Physics Library, D. Reidel Publishing Company.

Many people have contributed to fashioning me into a space scientist. There were my parents and school and university teachers who taught me to think. Ramachandra Rao and Bill Bartley assisted in the design and construction of the Pioneer and IMP instruments. My PhD research supervisor, Dr Geoff Fenton in Tasmania provided me with the resources and guidance that started my career as a space scientist. Many more marvelous friends assisted me with hard work, data, ideas and criticism.

I acknowledge with gratitude the assistance and guidance of New Holland Publishers, and my editor, Michael McGrath.

Finally, my wife Gillian has supported and participated in my adventures in space science since soon after the launch of *Sputnik* and assisted me mightily in the completion of this narrative.

Foreword

Fifty years ago few Australians grasped that space technologies would become a crucial component of modern society. Only a few knew the work of the Romanian, Hermann Oberth, in the 1920s and his vision of an orbiting space station, or about Robert Goddard's rocket experiments in the United States.

Arthur C Clarke's 1945 predictions—that satellites in geostationary orbit could relay radio signals from one side of the world to the other—were regarded as only of theoretical interest. Certainly, no one thought of a career in space science or technologies when my generation went to university in the early 1950s.

This is the world that Ken McCracken describes here. In four short years he went from being a research student in Australia's smallest university to building pioneering spacecraft that went to the orbits of Venus and Mars. Everyone was learning on the job—space science was being developed largely by trial and error. Soon he was one of a small group of scientists choosing experiments to be flown in the NASA space program. His story demonstrates the quality of Australia's education system and our national characteristic of 'giving it a go'.

Many years later, I asked Ken to give one of the lectures at a symposium I organised on Space Science as Minister for Science in the Hawke government. I asked others to speak on communication, meteorological and remote-sensing satellites. Ken was invited to speak on 'the lesser known, presently less important applications of space'. He spent most of his half hour talking about something called the 'Global Positioning System'.

No satellites had yet been launched; few in the audience knew anything about it and some said that Ken (well known for his enthusiastic approach to everything) had 'gone overboard this time'. Twenty years down the track and we now have GPS systems in transport and communication systems generally. This illustrates how rapidly the space technologies have developed to provide key elements of our modern social infrastructure that few believed possible only a generation ago.

The author stresses that he has not written a history of space research. Ken's book shows how individuals adapt when faced with ideas, actions and decisions that are far beyond what we learn during formal education.

Future generations of young Australians will venture into the unknown in the same way, in biotechnology, nano-systems, sustainable energy and in other technologies as yet unknown. It is to them that the author is speaking as he recounts his experiences in those uncertain days at the beginning of the space age. His book aims to give an insight into the uncertainties, fears, opposition and exhilarating moments that they will experience.

Ken McCracken has contributed greatly to Australia and this book is a gift to be savoured.

Hon. Prof. Barry Jones, AO
Professorial Fellow
University of Melbourne

Contents

Prologue

'*Pioneer*, go for launch.'

'Thirty seconds and counting.'

It was 2.35 am on 16 December 1965. We stood shivering on the road that led across the marshes to the launch pads at Cape Canaveral, Florida. We were listening to the countdown on our car radio.

'Twenty seconds and counting.'

The *Thor Delta* rocket stood brightly illuminated about three kilometres away. It looked so small.

We were very tense. The positions of the Sun and Moon dictated that we had a launch window of only 12 minutes on each of three days in which we could launch the *Pioneer* satellite. We had already used one of those days 24 hours ago, when the count stopped at T minus ten minutes and never started again.

The tension was palpable. In those days of unreliable rockets, would three years of work end up in the Atlantic?

If we made it into orbit, would our cosmic ray instruments work? We had fought the bureaucracy for the past two years. We were using electronic components that NASA had initially insisted were unacceptable for space flight. We had broken quite a few of the rules. If our instrument failed, we would be in deep shit. It could be the end of a very short career in space research.

‘Ten. Nine. Eight. Seven. Six. Five. Ignition.’

I had never seen a rocket launched at night before. I was totally unprepared for the intense brightness of the flames. Involuntarily I stepped backwards—and fell down the road embankment into a knee-deep pool of very cold marsh water.

I should have been elated that *Pioneer* was sailing serenely into the sky. Alas, my mind was on other matters. Having lived in Papua New Guinea and having seen what a crocodile could do when hungry, all I could think of were the alligators that lived in those swamps. My friends said I shot out of the marsh faster than *Pioneer.*

So I stood there; wet, scared and shivering—my career as a space scientist had really begun.

THE CHRISTIAN SCIENCE MONITOR, WEDNESDAY,

Compass Still Useful?

Magnetic Study in Space

By Robert C. Cowen
Natural Science Editor of The Christian Science Monitor

Space navigators of the future may find the old-fashioned magnetic compass a useful instrument.

Because of this, and for scientific purposes, Dr. K. G. McCracken of the Massachusetts Institute of Technology says mapping of the interplanetary magnetic fields is a first order of business for space scientists.

As regards short-term space flights, Dr. McCracken noted that "it is clear that for short trips in lightly shielded space vehicles, it would be advisable to undertake the trip when the chances of a dangerous exposure to solar cosmic radiation are small."

In particular, he explained it "would not be advisable to make such a trip if the interplanetary magnetic field lines led to a large, active sunspot, for it is in the vicinity of such sunspots that large solar flares invariably appear.

"If, however, the sunspot were near the center of the solar disk, the chances of a dangerous exposure would be reduced, since a magnetic 'road' would not connect the sunspot with the earth."

VOL. 9, NO. 77, RICHARDSON, TEXAS, THURSDAY, AUGUST 18, 1966 PRICE 5 CENTS

Pioneer VII Satellite Carries GRC Cosmic Ray Detector

X-ray stars detected by Tasmanian equipment?

UNIVERSITY of Tasmania scientists are almost certain that galactic X-ray equipment assembled and tested by them has recorded the first X-ray stars ever to be detected deep in the Southern Hemisphere.

Success of huge balloon

CANBERRA. — A huge balloon carrying 382lb. of research equipment travelled more than 200 miles from Waikerie (S.A.) to Balranald (N.S.W.) in 6¼ hours yesterday.

The helium-filled balloon, 260ft. high and 176ft. in diameter, was part of an experiment to pin-point the source of X-rays emanating from outer space.

Satellite system key to water control improvement

An upgrading of Australia's Landsat resource-satellite system would mean dramatic improvements in the management of Australia's scarce water resources, fragile range lands and agricultural resources, the chief of the CSIRO Division of Mineral Physics, Dr Ken McCracken, told the National Science Forum in Canberra yesterday.

"With our small population and large land area, we Australians stand to benefit from this technology more than virtually any other country in the world," he said.

The CSIRO's Dr Ken McCracken made an equally strong plea for Australia to become involved in space but backed it up with figures and examples from experience. McCracken's division of mineral physics pioneered the techniques used in enhancing and interpreting data from the US Landsat series which have pinpointed everything from the huge Roxby Downs mineralisation to isolated, illegal plantations of marijuana.

McCracken identified the major Australian satellite needs as communications (Aussat) and resources management.

Space prospecting pioneers

By science and technology writer JULIAN CRIBB

RAINBOW spectra of a thousand hues sparkling from the impassive earth to expose its mineral secrets have propelled Australia to a world technology lead and yesterday led to the awarding of the nation's most prestigious science prize.

Three Australians and an American will receive the $300,000 Australia Prize for their work in remote sensing of minerals and the environment, the Minister for Science, Senator Cook, announced in Sydney.

Maneuvering By Pioneer 6 Sets Course

PASADENA, Calif. (AP) — The Pioneer 6 interplanetary satellite, its position oriented for maximum performance, sped on its 50-million-mile mission around the sun today.

32 Part I—THURS., DEC. 17, 1964 Los Angeles Times 2★

Science Preparing for Study of Space Energy

Solar Activity Slight for Next Two Years, Allowing Probes of Intergalactic Forces

BY GEORGE GETZE
Times Science Writer

AUSTIN, Tex. — In the next two years science will get its best chance yet to study the energies of intergalactic space that pour through the Solar System, beating against Earth like water against a rock.

According to Kenneth C. McCracken, of the Earth and Planetary Sciences Laboratory of the Graduate Research Center of the Southwest, it will be another 22 years before as good an opportunity comes along.

Sun and Beyond

Many scientists are taking advantage of these "years of the quiet sun" to study the sun itself. McCracken and his associates at the research center in Dallas, W. C. Bartley and U. Ramachandra Rao, are using the quiescent period to study what is beyond the sun and what is normally obscured by the tremendous radiations of the

‹ 1 ›

Dawn of the Space Age

Nowadays satellites are accepted as a normal part of life. It is no longer remarkable to see images of the other planets in our solar system. However, 50 years ago none of this existed. Furthermore, even the most far-sighted engineer or scientist totally underestimated the progress that would occur in ten, twenty and fifty years.

On 4 October 1957 the Soviet Union launched an 84-kilogram (185 lb) satellite into orbit. They named it *Sputnik*, 'little moon'. A month later they launched a 500-kilogram (1100 lbs) satellite, with the dog Laika as a passenger. Several months later America launched *Explorer 1*, and discovered the Earth's radiation belts. Mankind had entered the Space Age.

At the time I was a research student at the University of Tasmania and was conducting what in those days passed for 'space research'. I recall that my first—and slightly unworthy—reaction was that the satellites would render my work useless and spoil all my fun. How wrong I was. Just two years later, I decided 'if you can't beat them, join them'.

Nowadays satellites are accepted as a normal part of life. Each night the weather report on television shows an image of Australia taken from a height of 25,000 kilometres (16,000 miles). Ships, farmers, bushwalkers and many cars use a global positioning system (GPS) to tell them where they are. The pubs and clubs of Australia all have small, parabolic antennae on their roofs, which receive cricket, football and other forms of entertainment relayed from communication satellites located far above us. We track the flight of albatross and the wanderings of seals, whales and polar bears using satellites too.

It is no longer remarkable to see images of the other planets in our solar system. We have seen the rings of Saturn close up, the impact of the comet Schumacher-Levy on Jupiter, and televised views of the outermost planet, Pluto, in 'real time', as the techno-freaks call it. Using the World Wide Web, we have observed the daily travels of the mobile laboratories *Spirit* and *Opportunity* as they moved around on the surface of Mars.

Our astronomers have observatories orbiting in space, which allow them to see the stars and the universe using wavelengths that cannot penetrate the Earth's atmosphere. The gamma rays, X-rays, ultraviolet and infra-red light and microwave radiation that they see are revolutionising our understanding of the universe.

All of this now seems so normal, so obviously right and proper, and maybe so boring, unless you are a 'nerd' or into technology. The youngest child knows about rockets, satellites and even inter-galactic travel.

Fifty years ago, however, none of this existed. Even the most far-sighted engineers or scientists totally underestimated the progress that would occur in ten, 20 and 50 years. To a casual observer today, all these innovations and improvements have come to us in a great, free-flowing stream; we don't question their origins.

This seemingly serene progress from one technological marvel to the next is rather like a swan swimming across a calm pond. On the surface the swan sails calmly forwards, a nice little bow wave adding to the picture of poetry in motion. Under the water its legs are thrashing back and forth. Down there, it's chaos.

On the surface the swan sails calmly forwards …
Under the water … it's chaos.

This book is not a history of space exploration and technology. It does not tell you the hows, whys and wherefores of the Space Age, as we now know it.

Its role is quite different. It gives the personal experiences of a young Australian scientist who just happened to be in the right place, at the right time, to participate in the early, highly uncertain days of the dawn of the Space Age. This book outlines some of the crazy ideas that drove my colleagues and I as space pioneers when no one knew any better. It lifts the veil on some of the stuff-ups that we experienced. And it touches on the human side of space exploration—the macabre sense of humour, the jokes and the personalities of those early days. It tells you about the chaos underneath the apparently smooth advances in science and engineering.

I turn the clock back to 1956, the year before *Sputnik*, and reflect on what we knew then about the potential of rockets.

Guy Fawkes Day, more generally called Cracker Night, celebrated an attempt to blow up the British Houses of Parliament on the 5 November 1605. It was one of the high points of a young child's year, and for weeks neighbourhood kids would collect garden rubbish and prepare a bonfire on a vacant lot for the great night. The bonfire would be lit and as many crackers and fireworks as one could afford would be exploded in rapid time.

The skyrocket was a firm favourite, and children loved to fire them, often up to 30 metres (100 feet) or more. However, in Australia in 1956, few were imagining that rockets could open up the horizons of the universe.

Some people did know that there had been considerable discussion of space travel and space stations in the magazines of the 1920s. Then the Nazi's *V2* Rocket brought home the destructive potential of rockets—V stood for *Vergeltung* (retaliation). These were the large rockets, with an explosive warhead of around 1000 kilograms (2200 lbs), that the Germans had aimed at London during World War II from launching sites in France and Belgium. After firing the rocket climbed to a height of almost 100 kilometres (60 miles), then descended at supersonic speeds, so that you only heard it coming after it had exploded, creating the added terror of uncertainty for the British civilian population.

The German engineers who built the *V2* also designed the *V4*, which would allow them to launch a satellite into orbit around the Earth. The odd physics professor or communications engineer in Australia understood the implications: such a satellite could be used for spying or long-distance radio communications. Most people took the view that perfectly adequate ways to do this existed already and they regarded the idea as a bit nutty.

The only people who took the possibilities demonstrated by the *V2* and the plans for the *V4* seriously were the military planners of America, Russia and several European nations. To them this was 'top secret stuff' and they weren't talking.

To the man in the street, space travel and satellites remained the stuff of fantasy. Like Goldilocks, Cinderella or the exploits of Buck Rogers, they were something that was too fanciful to be true.

Mind you, the prevailing attitude towards science in general was little better. When I started high school in 1945, the community had little understanding of or interest in science, other than as a subject that was inflicted on some kids in class. Some of us pre-teenagers had been exposed to remarkably good science by the Australian Broadcasting Corporation's (ABC) *Argonauts' Club*, which implanted undefinable ideas and questions in our young impressionable minds that later led some of us to become the scientists and engineers of the next generation.

Apart from that, science then was practised by the 'mad scientist' who appeared in the comics of the day—it was widely believed that there was no other kind.

Science then was practised by the 'mad scientist' who appeared in the comics of the day—it was widely believed that there was no other kind.

That may seem surprising now. The fruits of 19th- and early 20th-century scientific discoveries already surrounded society by the early 1940s. All homes in the cities had electricity, a photographic camera and radio (then a 'wireless'). However, each of these advances were said to be due to the work of 'an inventor'.

That the inventor was often an engineer or scientist was totally unrecognised by the public in general.

There were exceptions, of course. Albert Einstein was a darling of the intelligentsia. The better newspapers would occasionally run an article on his theory of relativity. However, since even the average science teacher then had only the vaguest idea what it was all about, these articles probably did more to reinforce the image of the 'mad scientist' than to educate.

The bright young student did have some knowledge of Einstein—often for the wrong reason. Thus as I entered my senior years in high school, I knew the doggerel:

There was a young fellow named Bright,
Who travelled much faster than light,
He started one day,
In a relativistic way,
And ended up the previous night.

In mathematical terms this is written as:

$t^1 = t/(1\text{-}v^2/c^2)^{0.5}$

… and says, roughly, that the clocks on a space ship will seem to run a little slower than those on Earth.

There was also its companion:

There was a young fencer named Fisk,
Whose actions were incredibly brisk,
So fast were his actions,
The Fitzgerald contraction,
Made his rapier appear as a disk.

That is:

$x^1 = x / (1-v^2/c^2)^{0.5}$

... which says, roughly again, that the length of something in a space ship appears different to someone looking at it from Earth, than it does to someone on the space ship. Scientists and university students soon saw that these rhymes had interesting potential. Thus the second rhyme became:

There was a young physicist named Fisk,
Who in bed was exceedingly brisk ...

... and I leave it to the reader's imagination to decide which part of Fisk's anatomy appeared like a disk.

The symbol 'c' in these mathematical equations stands for the velocity of light, which is 300,000 kilometres (186,000 miles) per second. So the very idea that anything could move at this speed was seen to be completely 'ratbag' stuff. Once again this did nothing to dispel the mad scientist image.

In a material sense, few of the fruits of science and engineering that we now take for granted were in evidence then. Telephones were relatively uncommon, and interstate calls were a rarity. The first seeds of the electronic age were present—in the form of the household wireless. It used electronic valves, which usually looked rather like an elongated version of a light bulb, glowed brightly and broke down with monotonous regularity. However, they enabled things to be done in the home and in the laboratory that had been quite impossible just ten years previously.

In 1956 Australia was still a very regional society. Telephones, radios, aeroplanes and other forms of technology could only service an average area of 200 kilometres (124 miles) from reasonable sized towns. The technology did not yet exist to have nationwide radio broadcasts, as we know them now.

The adventurous soul, desiring to hear news or programs from afar, would purchase a 'short-wave radio'. This required you to string a long piece of wire between two 15-metre (50 ft) high trees in order to 'pick up the signal' and hear a squawky, noisy voice from afar—it was exciting and wondrous.

Television was unknown. Computers, as we now think of them, were decades in the future. Lasers, transistors and all other 'semiconductors' were yet to be invented. Antibiotics were unknown to the general public. The plague diseases were still very real to us all—friends were crippled by poliomyelitis (polio) or died of tuberculosis, and I myself spent many months in bed with recurrent bouts of bronchitis. Surgery was limited to things that we go to hospital for only a day or so now. If you needed anything else … well, bad luck.

Science was seldom seen as a career. Professions like engineering or industrial chemistry were good if you were bright enough—and a 'swot'.

Beyond that, the concept of a life in research hardly existed. At university the few who saw research as their future accepted that they would earn very little money or recognition. Scientific research was an avocation—you did it because you wanted to do it—not because you would be well-off.

My father was a Scot. The Scottish were big on learning. They often became the mechanics, the engineers, the surveyors and the university professors of the British Empire, as it was then. His family being poor, he left school at the end of primary school at the age of twelve, and gained employment as a 'measurer' in a steel mill. As with most young people then, he attended 'night school' and obtained accreditation as an accountant at the age of 37. He became one of the most senior men in the Australian Taxation Office, and was destined for a higher position before he died at the age of 56. You could go a long way with a primary education and determination back then.

With this background he was determined that I should take maximum advantage of the excellent education that was available in the state schools of Australia. I was encouraged to be numerate, to read a lot and to do creative things. The parental expectation was that I would be a technical person of some kind—an actuary perhaps, or a chemical engineer or the like. To this end, it was always suggested that I would go on to university one day. This was an unusual expectation for a child at that time, when the majority of children left school before the end of their third year at high school.

Clearly, I was very lucky.

‹ 2 ›

The first stirrings of science

At the age of eleven, I found a description of how a radio transmitter works. It gave detailed drawings of the component parts. Now here was a real challenge.

I always liked making things. First I used a big Meccano set. This consisted of numerous strips of metal, approximately one centimetre (½ inch) wide, with holes drilled down their length, plus wheels, nuts and bolts. The instruction book showed you how to build motor cars, trains, cranes, forts and all the things that interested children in those days. Countless children of my era used them and grew up to become the engineers, nuclear physicists and mechanics of the world. With these I would build models of many things, then pull them apart to make something else.

This was followed by a major obsession in building billy carts. The billy cart was a wooden box, onto which wheels—generally from an old pram—and a steering mechanism were attached. We would find a street on the side of a hill and careen down it, oblivious to the traffic.

This risk to life and limb was only an incidental pleasure, though. The real pleasure was the process of creation. Having built a cart, plans were then made to build a better one. The existing one was torn apart; a few new pieces of timber were liberated from adult use; and the advanced model was built and tested—and so on.

At the age of ten I graduated to model aeroplanes. There were no prepackaged kits and plans in those days. When you were beginning, you used the designs provided in the comics of the day. Soon you drew your own design, then built it. There were failures, but with time I was building sailplanes with a wingspan of a metre and more.

Then I graduated to chemistry. My father presented me with a second-hand chemistry set, which included an instruction manual for various simple experiments. Making solutions change colour soon lost its attraction. However, a bit of research told me how to make explosives.

I soon mastered gunpowder and another marvellously violent explosive based upon sugar. I would pack this into the brass cases from machine gun ammunition—commonly available to the adventurous child during World War II. This would be sealed with toilet paper and a piece of dynamite fuse to set it off. It does not take much imagination to see that this would be quite dangerous—it was. Somehow I didn't injure anyone.

I also learned about electricity. I used my chemistry set to make batteries. I made electromagnets, electroplated metals and learned about switches, circuits and so on.

Then, at the age of eleven, I found a description of how a radio transmitter works. It gave detailed drawings of the component parts. Now here was a real challenge.

The description was in a children's book published in 1916 and was the design of a 'spark transmitter'. Such a device would have been familiar to Guglielmo Marconi, the Italian engineer who proved that wireless technology would work in the 1890s.

It never occurred to me that the technology might have changed. So I wound coils of wire. I made Leyden jars by lining the inside of jam jars with silver paper saved from blocks of chocolate. I powered it from the 240-volt mains. How I did not electrocute myself I will never know.

Such a device would have been familiar to Guglielmo Marconi, the Italian engineer who proved that wireless technology would work in the 1890s. It never occurred to me that the technology might have changed.

My uncle, who was in the navy, had given me a morse code key some years previously. This was a special form of electrical switch, which allowed you to make the dots and dashes of morse code. The letter 'A' is *dot dash*; the letter 'Z' is *dash dash dot dot.*

In those days nearly all long-distance communication was done by morse code, and it never occurred to us that this would ever change. Morse code was one of the pillars of modern technology in the 1940s, and learning it was considered to be a very good thing.

I built the morse key into the transmitter, which gave it a nice grown-up look. I taught myself to send morse code by tapping out parts of the popular children's book, *The House at Pooh Corner*, by A A Milne.

The 1916 book had explained that a radio transmitter sends its signals out into the 'aether' from an antenna. I strung a long piece of wire to a small tower I built in the garden using tomato stakes. I connected my transmitter to it and tapped out more bits of *Pooh Corner.* My mother tried to tune me in on the household wireless set.

Adjustments were of no avail. Nothing was heard and I assumed that the experiment had failed. I was always very philosophical about failure, and now know that this was a marvellous training for my later life in science.

It is just as well that the experiment failed. It had never occurred to me that governments made rules about radio transmitters. Worse still, this was in the darkest days of World War II. Fierce fighting was in progress on the Kokoda Track. Darwin had been bombed; Sydney attacked by three submarines; and a number of other cities had been shelled. All radio transmitters had been confiscated years before. Only spies had them.

To this day, I wonder what would have happened if my transmitter had worked. What would the bureaucracy have done about the eleven-year-old son of a senior public servant, transmitting the immortal words of Winnie the Pooh, on a World War I spark transmitter, from a house opposite the United States Embassy in Canberra?

I almost wish that it had worked.

‹ 3 ›

Home-brew rockets

Guy Fawkes Night had demonstrated to me that sky rockets worked in the atmosphere. However, they were small and would only go to a height of 30 metres or so. The challenge was obvious—to make one that would go much higher.

I first learned about rockets, satellites and space travel from an unlikely source; a comic strip. One of my parents' friends had a stack of comics that their son had bought in the 1930s, and I was given them when I was confined to bed for a month with bronchitis. The hero was a space pilot called Buck Rogers, and he flew to the various planets of the solar system and fought battles with the baddies using 'annihilation rays'.

Then I learned from the ABC's children's program the *Argonauts Club* how aeroplanes and rockets worked. To this day I remember the difficulty I had in understanding how a rocket could work out in space. I had no problem when it was in the atmosphere, for I reasoned—wrongly—that the exhaust gases would push against the atmosphere. Out in space, though, it would have nothing to push against. My parents and their friends were of no assistance; I was not yet in high school, and it never occurred to me that there was such a thing as a science teacher. I wasn't convinced that a rocket would work in space at all, although several years later, after the German *V2* rockets started to land on London during World War II, we all realised differently. Perhaps, I thought at the time, the comic is wrong.

Guy Fawkes Night had demonstrated to me that sky rockets worked in the atmosphere. However, they were small, and would only go to a height of 30 metres (100 ft) or so. The challenge was obvious—to make one that would go much higher.

I mastered the manufacture of a powerful form of gunpowder and built several rockets with it. However, they were all dismal failures; they exploded violently and were the cause of some parental displeasure. I moved onto building wireless sets, and shelved the idea of a big rocket to some later time.

This opportunity came six years later. I had now graduated from high school, but was too young to go to university. To fill in a year, I was employed as the laboratory boy in the research department of the Electrolytic Zinc Company, located on the outskirts of Hobart. This exposed me to real scientific research at the tender age of sixteen. However, to me, it was also a marvellous opportunity to try out some of my own ideas.

An important part of my job was to look after the chemical store, and to order new chemicals when the scientific staff needed them. In reality this was a bit like employing a poacher as the gamekeeper. I experimented with my various passions—photography, hydroponics and, finally, my long latent desire to build a big rocket.

There were resources aplenty. I researched the technical details and concluded that my earlier experiments had failed because the gunpowder had burned too quickly. I used my limited knowledge to make a propellant that would burn more slowly.

The great day came. The launch facility was made from the construction material of last resort at the zinc works—25-kilogram (55 lbs) zinc ingots. It was located in a derelict factory building adjacent to the research department. The building had been used for about twenty years to process the fine, brown zinc concentrate, and every horizontal surface was covered with centimetres of the stuff. Computations show that there were tonnes of it on the roof girders. The building was totally open at one end and looked out onto the River Derwent. The wick was lit. We stepped back several metres for safety.

There was an enormous explosion. Obviously the propellant was still burning too quickly. A 25-kilogram zinc ingot landed at my feet. Two seconds later a great, choking cloud of zinc concentrate fell on us. An enormous cloud of the stuff roiled out the end of the building and down the hill towards the River Derwent. I was told that from a distance it appeared as if several factory buildings had exploded. The company fire brigade arrived within minutes.

I was told that from a distance it appeared as if several factory buildings had exploded. The company fire brigade arrived within minutes.

There were two remarkable outcomes of this little experiment. First, I was not fired on the spot. Second, and more remarkably, absolutely nothing was said to me about it by my boss or anyone else in authority. It seems that the view in those far distant days was: boys will be boys.

‹ 4 ›

University days

Using parts from old radio sets and junk thrown out by radio shops or liberated from the army, I built a transmitter for virtually no cost while still a teenager.

I started university in 1951 with the intention of becoming a chemical engineer. Chemistry was still my greatest interest. It seemed a good way to continue that interest and to find a good job. My father, true to his Scottish roots, was pleased with all of this. He liked the idea of science and engineering, since he saw them as 'important to society'.

However, mischief was afoot. I had won several prizes in my last year of high school. As was common then, they consisted of a sum of money to allow me to buy textbooks of my choice. One that I bought was a very comprehensive encyclopedia of nuclear science, at a level I could understand.

Here was the story of how the complexities and order of the atom and its nucleus had been unravelled, often by people only a few years older than I. I was astounded that I could understand a lot of it. I was hooked. I wanted to be a physicist—perhaps even a nuclear physicist.

This worried my father a great deal. The witch-hunts of the McCarthy era in America were in full swing. Communists and communist sympathisers were the main target, but people with liberal and humanistic views were caught up in them as well.

The physicists who had participated in developing the atomic bomb were obvious targets and there were many cases where their liberal views were not appreciated by the government. Physicists were suspected of espionage and there were enough proven examples to put everyone under intense and sometimes unreasonable pressure. Closer to home there had been a major fuss about the possibility that the CSIR—as Australia's Commonwealth Scientific and Industrial Research Organisation (CSIRO) was then known—had been infiltrated by 'reds under the bed'. My father saw little attraction in allowing his son to enter such a profession.

He outlined the problem to me as he saw it. Then he said, 'It's up to you. Your mother and I will support you in whatever you decide'. I did change from chemistry to physics and mathematics, but his astute reading of the dangers that can occur in pioneering science remained clearly etched in my mind.

At the end of my first year at university in 1951, all the male students were herded off to military camps as the second intake of the Australian National Service scheme. This was a mild form of compulsory military training; three months of continuous training in an army camp near Hobart, followed by three years part time.

After a little negotiation I arranged to be transferred from the university regiment to the Australian Signal Corps. This gave me great scope to play with radio transmitters and receivers—from big brutes that caused the electric lights to flicker in sympathy with the morse code we were sending, to small ones that we carried on our backs. Weekend training exercises consisted of driving six to eight vehicles all over Tasmania and setting up our wirelesses each hour to talk to one another. Needless to say, it was fun and we were paid for it—tax free!

Many of the older members of the signals company were radio amateurs—this meant that they were allowed to build and use transmitters in their homes to talk to other amateurs anywhere in the world. As we toured Tasmania on our weekend exercises, they would occasionally tune the transmitters from the army frequencies to those used by the amateurs, and play at amateur radio as well. This was such an obvious extension to my long-standing interest in radio that I quickly set out to become a radio amateur as well.

These days it is hard to understand that radio was regarded as an exotic activity in the 1950s, and it fascinated many. However, consider: in 1953 my parents did not have a telephone in our house. I used a public telephone perhaps once a year to make a local call. Interstate phone calls were rare, and overseas calls almost unheard of.

By the 1940s radios had become a central feature of family life. Families would listen to the radio over dinner. The radio serials such as *Dad and Dave*, *Yes, What?*, *Much-Binding-in-the-Marsh*, *The Lawsons* and *Blue Hills* went on for years, as did the ABC's *Argonauts*. Radios, however, tended to break down fairly often,

There was a real sense of romance and achievement in being able to communicate with people throughout the world using a transmitter and receiver that you had designed and built yourself.

As a consequence the radio repair shop became an important part of every town. They were also a very important 'resource' for children like me. Their rubbish bins were always worth raiding to find radio valves and other bits and pieces. The proprietors, often radio amateurs themselves, were always helpful and often generous.

So using parts from old radio sets and junk thrown out by radio shops or liberated from the army, I built a transmitter for virtually no cost while still a teenager. The greater difficulty was gaining the licence to allow me to transmit to the world. Officialdom decreed that you must pass a technical exam, and to be able to send and receive morse code quickly.

Those exams passed, I was at liberty to talk to other radio amateurs—in America, Canada, Asia and Europe—all with a little transmitter with the power of only a few watts. This was magic indeed.

There was a real sense of romance and achievement in being able to communicate with people throughout the world using a transmitter and receiver that you had designed and built yourself. To this day I vividly recall the feeling of wonder and exhilaration when I used my first set, built from the small 6V6 radio valve of an old home wireless thrown out by a radio repair shop, to hold a two-way conversation with another amateur in Casablanca.

So, to my first encounter with the policemen of the airwaves.

This initial run-in was minor. Our house was situated close to the Hobart Roman Catholic Cathedral. Being technologically alert, the priests had installed a public address system, which would help them to keep everyone awake.

At first there were simply repetitive clicks, to a rhythm of sorts, but hardly noticeable in those days of lo-fi (low-fidelity amplifiers and speakers) and the Latin mass.

I then graduated from using morse code to voice transmissions. In the middle of an invocation, the congregation was startled to hear, 'See-cue DX, see-cue DX, this is Victor Kilo seven Kilo Mike', repeated over and over. Translated from amateur jargon, that meant that I wanted to talk to anyone a long way from Australia. Heaven, however, was not the immediate intention.

In the middle of an invocation, the congregation was startled to hear, 'See-cue DX, see-cue DX, this is Victor Kilo seven Kilo Mike', repeated over and over.

Investigations were made into the source of the interference. The problem was found to be in the electrical cables in the public address system in the cathedral. However, from that time, my file started to grow.

It filled out rapidly a year or so later.

At the university my work was part of the Australian Antarctic Research Program. Many of my friends would disappear each summer to spend a year in some God-forsaken place such as Macquarie or Heard Islands or the Antarctic continent itself. The prospect of a year in the freezer seems to have had some strong hormonal effects. Each year, there would be a crop of sudden marriages shortly before the ships sailed south.

Now it is easy to telephone people in the Antarctic. In the 1950s it was totally impossible. The only official communications between the Australian bases and the rest of the world were by morse code. The Australian Post Office provided a telegram service and we all became skilled in saying a great deal in ten well-chosen words. For the Antarctic expeditioners, the Australian government—in a display of compassion and generosity—allowed a free 25-word telegram each month, to and from their loved ones.

Well, you can't really say much with 25 words, either. So codes were developed: WYSSR (pronounced whisser) meant 'I love you and miss you very much'. A string of 25 whisser-like words would be sent to Australia by morse code and mountains of pent-up emotion and loneliness would be expressed in the form of a scruffy piece of paper, which looked like a recycled brown paper bag.

Understandably, the expeditioners found this whisser business to be less than satisfactory.

Enter the radio amateurs. At this time they were happily chatting back and forth between Antarctica and Australia with real voices. Australian amateurs were not allowed to send messages for other people, but the idea soon arose that the amateur in Australia would just happen to be visited by the wife or girlfriend of an expeditioner at the time that the amateur was talking to Antarctica. A coincidence, of course.

So in the depths of winter, each Saturday night, yours truly would talk to the Australian Antarctic base on Macquarie Island. The new wife of one of the physicists on the island would visit me and join in the discussions of our scientific experiments and other topics of interest.

So far so good. However, the aurora australis would send waves of electrons back and forth across the ionosphere and divert our signals to other places on earth. Assorted Chinese gentlemen would transmit on the same frequency and the 100 strokes of lightning that occur worldwide each second, all seemed to be drowning out our voices. Conversation was not easy. Nor was it romantic to hear your loved one sounding like Donald Duck talking underwater.

Nor was it romantic to hear your loved one sounding like Donald Duck talking underwater.

Scientists refer to this as an example of one of the great unwritten laws of science—the inherent 'cussedness' of nature. However, in this case, the Antarctic radio amateur had a solution.

First let me digress. Radio amateurs are only allowed to use their transmitters to talk to one another, as I have said, and to perform experiments. One of the most interesting—and often very educational—types of experiment in the early 1950s was with antennae. These were much bigger than the antennae we have on our homes to receive television.

The antennae would be 20 to 40 metres (65–130 ft) long and ideally they needed to be 20 or more metres above the ground. Few homes have two 20-metre high trees in their backyard and, even if they did, they always seemed to be lined up in the wrong direction to send the radio waves where you wanted them to go. So we would experiment with ways around these problems. Wires would be draped from trees, poles, wherever. Some worked very well. Most did not. But much fun was had in building, trying out, then repeating the whole process. It was very common for someone to say, 'Stand by, I will change to another aerial, and will you please compare my new signal strength with my present signal'.

So back to our amateur on Macquarie Island. Frustrated by nature, he would say, 'Hang on a minute, and I will change to a new antenna'. A minute later his voice would boom through, loud, clear and emphatic.

To such a display of the might of science and the ingenuity of mankind, the right reply was 'Your signal is now very good', and drop the subject quick smart. Those not in the know would ask a dumb question like: 'What sort of antenna have you changed to—my signal strength meter tells me that your signal has increased by 20 decibels?' That is: impossibly good.

The grubby truth is that it was the same antenna. However, the amateur had changed from his small, homemade transmitter, to the big, powerful brute provided for official communication.

Back to frigid Hobart. At the agreed time two tiny radio signals tried to find one another. So Macquarie Island 'switched to a new antenna', and I did the same in Hobart. The effect was remarkable. I could be heard very well on Macquarie Island—and throughout half of Hobart on the normal radio broadcast band.

The public servants who were the guardians of the virginity of the ionosphere snapped into action. Tape recorders rolled. Names were looked up. Typewriters rattled.

As a result I received a nice little letter the following Monday, asking me whether I could think of any good reason why my amateur licence should not be cancelled and my equipment confiscated.

As a matter of fact, I could. It became more complicated when the eunuchs of the ionosphere realised that I did not own the transmitter I had been using. It belonged to the Australian Army, who seemed very unhappy about the idea of confiscation.

Bureaucrats are pragmatic people. They usually know when to quit. So that little episode died quietly, leaving its mark only in an obscure filing cabinet and in the minds of the eunuchs of the airwaves.

A year or so later, the eunuchs snapped into action again. The occasion was the annual university 'muck-up day'. This normally took the form of sporadic vandalism, such as one incident which involved putting a rowing boat in the park pond in the centre of the city. The fountain in the pond was then filled with detergent, so that the boat disappeared in a mountain of foam.

The university parade was the major event of the day. Nothing much of significance ever happened in Hobart, and the burghers of the city looked forward to this annual break from the monotony. The students would organise about fifteen floats, specialising in questionable taste and with often ribald commentary on the politics and personalities of Tasmania's biggest city. Hobart loved it.

The legal fraternity had always featured strongly in these events. On one occasion four legal students staged a mock robbery of the main branch of the Commonwealth Bank. Two of them ran out of the bank, clutching two Gladstone bags, and ran down the middle of the main street of Hobart. Thirty metres behind were two students dressed as security guards, firing toy pistols and shouting out 'stop thief'. The local constabulary was not amused, and the students all ended up in court.

It is interesting to note that one of the students later went into federal parliament and became a participant in the senate enquiry into the Australian banking system. I have always wondered if he included his banking escapade in his curriculum vitae.

Sooner or later the day had to come when the muck-up day went 'high tech'. Thus, when the news started on a Hobart radio station one balmy evening, there was a click and a slight pause, followed by a most unusual news broadcast, largely consisting of university humour of very poor taste.

The conspirators had realised that the radio transmitter was unattended on a hill some distance outside of Hobart. They made a preliminary visit to the transmitter and found that the technicians had very thoughtfully left a complete set of instructions and electrical diagrams.

The students then recorded their version of the news on a tape recorder and revisited the transmitter just prior to the commencement of the news. They disconnected the studio in Hobart and connected their tape-recorded 'news'.

The hill was quite isolated and the conspirators knew that it would take the police more than fifteen minutes to reach them. So after their fifteen-minute broadcast, they reconnected the line to the studio and slipped away leaving the hill along a walking track.

The establishment of Hobart was not amused. The local paper, the *Mercury*, thundered about irresponsible anarchists. The policemen of the airwaves were put on red alert.

Two days later the eunuchs trooped into my office at the university. No doubt they were thinking, 'We've got him this time', and planned to send me off to some horrible electronic gulag. To their enormous chagrin I was able to prove that my wife and I had been on our honeymoon in the mountains 300 kilometres (185 miles) from Hobart and that I had not been involved. They departed, extremely crestfallen, and were never able to establish who did what.

I am pleased that I did not know then what I was to learn some years later. The eunuchs were almost right when they came to my office. While I had had nothing to do with it, my office had been involved. During my absence, the spurious broadcast had been recorded in my office, using some of my equipment. The perpetrators went on to become pillars of society—the president of the Institution of Engineers Australia, a headmaster of a high school and a solicitor of high standing.

The bureaucrats knew none of this. It remained an unexplained mystery, but they always had a sneaking suspicion that somehow I had contrived to be in two places at once. My file grew fatter.

A few months later my wife and I left for America. For several years, however, the ionospheric watchdogs would phone the physics department of the university, asking for me whenever there was some unexplained radio transmission coming from the vicinity of Hobart.

Two decades later I revisited Hobart as a grand panjandrum of CSIRO, in order to make arrangements with the ionospheric bureaucrats regarding the reception of the radio signals from satellites as they went whizzing by. To my great amazement, they were the same bureaucrats who I had known in the late 1950s. They were savouring the fact that this time I wanted something from them—and they were going to make me work for it!

At the same time they seemed dejected about something. When the negotiations were complete, the more technical bureaucrat drew me aside and explained, 'We decided we would start the meeting by producing your file from the 1950s,' he said, 'however, we couldn't find it'. He speculated that head office in Melbourne had asked to see it.

Perhaps they are going to get me yet.

‹ 5 ›

Exploring the cosmos

Long before we spoke of 'space research', the cosmic ray group in Tasmania, and those elsewhere in the world, decided to concentrate on understanding this most improbable and unexpected property of 'empty space'.

A hundred years ago the scientific world was agog with excitement. Henri Becquerel discovered radioactivity in 1896 and the radioactive elements were found to emit rays that passed through metals and even lead. The scientific world named them the alpha, beta and gamma rays. Then careful experiments—on board manned hot air balloons—showed that there was another radiation, much more penetrating than any of the other three. The greatest surprise of all was that these rays were not coming from the Earth—they were coming from out in space. They were given the name 'cosmic rays'.

For almost 20 years the cosmic rays presented a major scientific enigma. What were they? Where did they come from? An eminent Nobel Prize laureate claimed that they were very energetic gamma rays—he called them 'the birth cries of the atoms'—until a young Italian upstart took his equipment to Eritrea and showed that the rays were deflected by the Earth's magnetic field. In so doing, Bruno Rossi, who would become my mentor and advocate, initiated the tortuous process that has led to today's, still imperfect, understanding of these extremely energetic protons and other elemental particles of matter.

Before I continue, I digress and talk briefly about magnetic fields, which allowed Bruno Rossi to make his discovery. They are vital to modern society and will keep recurring in my story.

Magnetic fields are everywhere in nature and in our technological society. The electric motors in our refrigerators, motor cars and power tools would not exist without them. The hard disks in our computers and the tapes in our tape recorders are totally dependent upon the magnetic properties of iron oxides and some other materials. The compasses that have guided mariners since the 16th century would not work if the Earth did not have a magnetic field.

Magnetic fields have many interesting and useful properties. I will, however, only discuss the one that is important to this story: their influence upon moving electrical charges such as the proton (positive charge) or the electron (negative charge). A magnetic field exerts a force on them, perversely at a right angle to the direction in which they are moving. As a consequence they are deflected and follow a curved path. A positively-charged particle curves one way and a negatively-charged particle curves the other way, while an uncharged particle or gamma ray is totally unaffected and keeps going straight ahead. Bruno Rossi's measurements in Eritrea showed that the cosmic rays were deflected in a westerly direction, telling him that they were positively charged particles of matter and not gamma rays.

Throughout my career, my satellite measurements and theoretical studies have dealt with the magnetic fields in the Sun, in interplanetary space and those associated with the Earth itself. By measuring the manner in which cosmic rays are forced to follow curved paths through space, we have learned a great deal about those magnetic fields, and how to use them to protect astronauts as well as the communication and other satellites that orbit the Earth.

In the late 1940s the University of Tasmania had decided to perform experiments similar to those conducted by Bruno Rossi in Eritrea to see if further information on the nature of the cosmic rays could be found by measuring their properties in the Southern Hemisphere. They decided to make their measurements in two locations in Tasmania: Hobart and Macquarie Island. Led by Dr Geoff Fenton, they developed robust and sophisticated equipment, which soon provided the answers they sought.

Having gained a science degree in 1953, I made a key decision that led to my career in space science. (Of course, there wasn't anything called space science then, and we certainly never thought about performing experiments on satellites. Then that was crazy stuff.) I decided to study for an honours degree in the Tasmanian cosmic ray group, and to measure the cosmic rays with instruments located on the surface of Earth.

Looking back now, I can see that nearly all the scientists who became the space scientists of the 1960s and 1970s had, like me, decided to study cosmic rays or something similar, long before the start of the space era. We all just fell into space research more or less by accident, after the first satellites were flown.

In 1953 cosmic radiation was interesting for several reasons. Where did it come from? How did the cosmic ray particles attain such prodigious energies—far greater than we could make them then—or now—with the biggest 'atom smasher'? Since they were so energetic, these particles were also great tools to smash up atoms with, and for seeing what rubbish came out. Apart from anything else—lots of people received Nobel Prizes for doing this sort of stuff!

In the decade or so before 1953, it had been shown that the number of cosmic rays reaching Earth varied from day to day, and year to year. Explosions on the Sun caused the number arriving to change two days later. How could this be? At the time 'space' was believed to be just that, containing nothing at all, but something out there was capable of stopping some of these extremely penetrating rays, which could pass through many centimetres of lead. Long before we spoke of 'space research', the cosmic ray group in Tasmania, and those elsewhere in the world, decided to concentrate on understanding this most improbable and unexpected property of 'empty space'.

I became the junior member of the group seeking to understand this mysterious force. Up until then, cosmic radiation had been measured on the surface of Earth using 'ionisation chambers' and 'meson telescopes'. The readings of both were adversely affected by the varying temperatures of the atmosphere, which introduced unwelcome uncertainty into the results. However, there had been a new development. In late 1954 my supervisor, Geoff Fenton, gave me a slim, roughly copied document from the University of Chicago. It described the construction of a new instrument, called the 'neutron monitor', which showed none of this unhelpful effect. Geoff said, 'I suggest you construct a neutron monitor as part of your PhD work'.

The neutron was discovered in 1932 and was the key to the development of the atom bomb. Using the new instrumentation developed for that purpose, John Simpson at the University of Chicago had devised a sensitive, accurate means to measure cosmic radiation. It used many tonnes of lead, to trap any cosmic ray neutron that came by, and great quantities of paraffin wax, to slow the neutrons down. Then there were 'proportional counters' (detector tubes), filled with a gas containing the element boron.

A neutron, on colliding with a boron atom, caused a small electrical pulse to come out of the detector. The pulse was very small; the voltages used were high (3000 volts); and there were many things that could go seriously wrong. Worldwide, they did. The fates were good to me though; the very considerable expertise of an electronic engineer in the cosmic ray group, and my experience as a radio amateur, meant that we had no trouble whatsoever. By early 1956 we had completed and 'de-bugged' the second neutron monitor to be operated outside the North American continent. I was also building four more, much bigger neutron monitors—bigger to increase the sensitivity—to be installed elsewhere.

The pulse was very small; the voltages used were high (3000 volts) and there were many things that could go seriously wrong.

By then the Hobart cosmic ray group was expanding its operations to better understand the mysterious force out in the cosmos. To do this, we established a number of cosmic ray observatories throughout the Southern Hemisphere. I established my own cosmic ray hut containing my second neutron monitor on the slopes of Mount Wellington, at whose foot the city of Hobart is situated (Plate 2a).

Some of the answers we sought required that we make measurements at a number of latitudes. So observatories and neutron monitors were established, at various times, in Lae (Papua New Guinea), Darwin, Brisbane, Hobart, Antarctica, on Mt Wellington and Macquarie Island. Other answers would come from measurements at different longitudes—and these were provided by observatories in Hobart, on Macquarie Island and at Mawson and Casey in Antarctica.

By 1959 the Tasmanian 'cosmic ray network' had became one of the largest and most effective in the world. For my part, I had become a very experienced experimenter, which was a great asset for the rest of my life in science.

In those days there was little money for scientific research. Little of the Hobart network would have been possible without the support of the Australian Antarctic Program and our ability to scrounge things other people didn't want. We made great use of the enormous amounts of electronic and other equipment that was left over from World War II.

For example, Geoff Fenton, the leader of the cosmic ray group, purchased 20,000 radio valves that most people regarded as useless—and were therefore exceedingly cheap. Henry Ford supposedly said that people could have the T model Ford in any colour they wanted, as long as it was black. This was a bit like our approach; we could use any radio valve in the electronics we wanted to build provided that it was a 7C7.

We used the bases from anti-aircraft guns to support our cosmic ray 'telescopes'. I used the altimeters from World War II aircraft to measure the atmospheric pressure. We were frequent customers of the 'war disposal stores', where there was lots of left-over equipment. Nobody knew what most of it did—and it was therefore very cheap.

We built things that were impossible, or too expensive, to buy. Geoff Fenton developed the ability to manufacture metre-long (3 ft) Geiger tubes to count the cosmic rays, at a time when the manufacture of Geiger counters was still a black art. I found creative ways to build parts of my neutron monitors with toilet paper and surgical bandages. This was still the era of 'do-it-yourself' science, and the Tasmanian team were experts at that fine art.

I found creative ways to build parts of my neutron monitors with toilet paper and surgical bandages.

Then, shortly before the launch of the first Earth satellite in 1957, nature provided a vital clue to the make up of the mysterious cosmic force that pervaded space. Very rarely, a great explosion on the Sun would launch a great burst of cosmic rays towards Earth.

The Tasmanian network was ideally placed to study these bursts and vital observations made in Tasmania and in America during two of these bursts in 1956 made it clear that there were magnetic fields in the solar system that were quite different from any of the magnetic fields known on Earth. No one had predicted these magnetic fields that, at the time, were completely inexplicable. Perhaps, it was argued, the mysterious cosmic force was associated with these inexplicable magnetic fields.

I developed a strong personal interest in these magnetic fields. I worked out a mathematical procedure that predicted how the cosmic rays would be deflected in these and in the Earth's magnetic fields. Without knowing it, I was launching myself into space research—a year before anyone knew there was such a thing.

‹ 6 ›

Meson song

From mesons all sorts of forces you get,
The infinite part you may simply forget,
The divergence is large, the divergence is small,
In meson field quanta there is no sense at all.

As a young scientist learns his or her trade, they learn how pioneering research is fraught with uncertainty, confusion and ridiculous notions. I was learning this fundamental truth in 1954 when I saw a poem that had been recited at a very important conference by Dr Edward Teller, the 'father' of the hydrogen bomb. He was a leading theoretician of his day and certainly understood the tortuous process of research.

This poem summarised well the complete state of chaos in nuclear physics in 1949. It could be altered, without the slightest difficulty, to describe any newly emerging field of science today.

Teller's poem was about the 'meson', a sub-atomic particle, which is heavier than an electron and about a tenth of the mass of the hydrogen atom. Two types of meson are seen in the Earth's atmosphere—if you are quick—they only live for a millionth of a second or less. At the time the poem was written, they had usually been seen as streaky lines on 'plates' (photographic emulsions) that had been flown to high altitudes on balloons or had been photographed in a device called a 'cloud chamber'.

The poem was originally addressed to nuclear physicists, so there are some technical terms that I will explain at the end of the poem.

First, though, a little context. Following World War II the new electronics and the high-flying balloons developed for the Cold War were used to study the properties of the atom and its nucleus. This research could also be done on the tops of high mountains, and many 'cosmic ray observatories' were established for this purpose.

Many scientists are keen bushwalkers, skiers and mountaineers, so it will be no surprise that many of these observatories were located to offer great scope for such activities. One such location was Echo Lake, high in the mountains of Colorado, USA, and one of the most important conferences of that time was held there in 1949.

History records that Edward Teller composed the poem in Central City, an old silver-mining town in Colorado, then recited it at the banquet of the Echo Lake Conference on Cosmic Rays:

> There are mesons pi, and mesons mu,
> The former ones serve as nuclear glue,
> There are mesons tau—or so we suspect—

And many more mesons we cannot detect,
Can't you see them at all?
Well hardly at all,
For their ranges are short and their lifetimes are small.

The mass may be small, and the mass may be large,
There may be a positive or negative charge,
And some mesons will never show on a plate,
For their charge is zero, while their mass is quite great.
What no charge at all?
No, no charge at all,
Or, if there's some charge, it's exceedingly small.

There are meson lambda at the end of our list,
Which are hard to detect and are easily missed,
In cosmic ray showers they live and they die,
However, you can't take a picture—they are camera shy.
Well do they exist?
Or don't they exist?
They are on our list, but are easily missed.

From mesons all sorts of forces you get,
The infinite part you may simply forget,
The divergence is large, the divergence is small,
In meson field quanta there is no sense at all.
What no sense at all?
No, no sense at all,
Or, if there's some sense, it's exceedingly small.

The last stanza of the poem is the most cutting and the most accurate description of the confusion that theoreticians can work themselves into.

For example, ' ... the infinite part you may simply forget' refers to the manner in which theoreticians often show that there are several 'solutions' to their equations. Frequently one of the solutions is 'infinity', a number bigger than the biggest number you can think of. The theoretician will then ignore this one and choose some much smaller 'solution', usually for no good reason except that it gives 'an interesting' result. Interesting it may be, but who's to say if it's right.

'The divergence is large, the divergence is small...' is the most cutting comment of all. The 'divergence' is a mathematical concept that, depending on the situation, may be *exactly* zero or *exactly* equal to some well-known number. The idea that it may be anything whatsoever indicates a degree of uncertainty that was very unsettling to the physicists of that time.

‹ 7 ›

Explosions on the Sun

In the first half of the 20th century, it was discovered that solar flares and magnetic storms greatly affected the ionosphere, which allows 'short-wave' radio waves to travel around the curve of the Earth. With the ionosphere out to lunch ... The commercial and political implications were great.

Having decided to study cosmic radiation, I quickly became acquainted with some of the eccentricities of our Sun. My subsequent career revolved around those eccentricities, and learning how to avoid some of their more unpleasant consequences.

Over 2000 years ago the Chinese observed that on occasions there were black markings on the Sun, which moved across its face in about fourteen days (Plate 5a). The remarkable scientist Galilei Galileo also observed these black markings with a telescope in 1609 and used them to study the rotation of the Sun about its axis. In so doing, he suffered the consequences of being politically incorrect. His discovery disagreed with the church's view that the Sun was pure and without blemish. For this and other 'misdemeanours' Galileo was investigated by his ecclesiastical superiors and house arrest followed.

Over 2000 years ago the Chinese observed that on occasions there were black markings on the Sun, which moved across its face in about fourteen days.

Nevertheless, it soon became very fashionable to observe the black markings, which became known as 'sunspots'. We in the modern world owe a great debt to the Chinese astronomers, Galileo and the others who started observing them from early in the 17th century. Their meticulous records show that our Sun has changed greatly over the centuries, with many implications for our modern satellites and technologies and for our understanding of how the Earth's climate has changed over the past five hundred years. We will return to these in later chapters. (Plate 5b)

Fast forward from Galileo to 1859. A 'gentleman scientist', later elected to the peerage as Lord Carrington, had established an observatory in London to study the Sun and map the sunspots each day. His family had done very well out of brewing beer, and he was not short of a bob or two. His was a well set up observatory and he could look at a large, clear image of the Sun and examine it in detail.

When observing near midday on 1 September 1859, he saw a bright sinuous line of white light develop, which was entwined with an extremely large group of sunspots. At first he thought that sunlight was infiltrating his darkened observatory through a hole in the roof. Having checked that, and seeing that the sinuous thread was growing still brighter, he left his telescope to find his assistant to act as a witness. When they returned several minutes later, the bright white thread was still there, but rapidly diminishing in brightness. They quickly sketched what they had seen.

The following day, Carrington visited the magnetic observatory that had been set up in London to assist mariners with accurate navigation by compass when far from land. He saw that the Earth's magnetic field had gone absolutely mad about seventeen hours after he saw the bright white line. He learned that another gentleman scientist, Hodgson, observing from elsewhere in London, had seen the bright white line at the same time. Worldwide there had been spectacular displays of aurorae, the northern and southern lights.

Carrington and Hodgson published their observations in the leading astronomical journal of the day. However, no one took much notice of them for the next 40 years, probably because no other examples were seen and these were merely 'amateur scientists', regarded with suspicion by the scientific heavies of the day.

Carrington and Hodgson published their observations in the leading astronomical journal of the day. However, no one took much notice of them for the next 40 years …

Then new technology was developed that allowed astronomers to look at the Sun in a very narrow part of the optical spectrum. Immediately, short-lived, bright emmission regions were frequently seen near sunspots. The observations of Carrington and Hodgson were vindicated. The bright regions were named solar flares. Frequently the Earth's magnetic field went mad about a day after a big solar flare; these magnetic disturbances were called magnetic storms. It became clear that Carrington and Hodgson had seen an absolutely enormous solar flare, which had been so intense that it could be seen against the very bright sun in white light.

Studies in the 19th century showed that the number of sunspots changed greatly from year to year in an eleven-year cycle. There would be very few for a couple of years; then the number would rapidly increase until the Sun appeared to have a bad case of acne; then the number would decline until there were no spots again about eleven years after the initial 'sunspot minimum'. Solar flares and magnetic storms were found to be more frequent in the years when sunspots were most prevalent.

Then, in the first half of the 20th century, it was discovered that solar flares and magnetic storms greatly affected the ionosphere, which allows 'short-wave' radio waves to travel around the curve of the Earth. With the ionosphere out to lunch, much of the world's radio communications, which in those days were largely by 'short-wave' radio, would be shut down for a day or so after a big solar flare. The commercial and political implications were great. Civilisation, as we knew it then, demanded that we learn more about the foibles of the Sun and how to avoid the inconveniences that it produced for us.

A large worldwide investigation of the Sun, and its effects upon the Earth was scheduled to coincide with the high level of spottiness that was predicted to occur in 1958. It formed part of the International Geophysical Year, in which scientists would carefully measure properties of the Earth's magnetic field, the ionosphere, the aurorae, the atmosphere and cosmic radiation to find out how all of these were influenced by the Sun and by one another.

Australia decided to measure all of these and more, all the way from near the equator to several locations in Antarctica. The Hobart cosmic ray group was going to establish observatories in Antarctica, Tasmania and Papua New Guinea. I started my research career in 1954, and, with the spotty year of 1958 looming and the prospect of all that data from many other places in the world, it was clear that the Sun would give me an interesting time.

I didn't have to wait long. My first research project had been to find out what was wrong with a cosmic ray instrument built for use at the Australian Antarctic base on Heard Island in the late 1940s. Called an ionisation chamber, it was meant to measure the effects of the cosmic rays on the atmosphere.

Once on Heard Island, it could not be made to work. Ionisation chambers were totally dependent on high levels of electrical resistance in certain key components, and the physicists on Heard Island, thinking that was the problem, had tried every trick in the book to increase the resistance. This included dissolving toothbrush handles in acetone to see if they would make better insulators. All to no avail.

The ionisation chamber ended up in disgrace in Hobart and I was given the problem to solve. This I did by brute force: I surrounded all the sensitive bits with a vacuum. It was then put into continuous operation, measuring the hour-to-hour and day-to-day changes in intensity of the cosmic radiation in Hobart.

It was a cantankerous beast, and something always seemed to be going wrong with it.

One day I went to check the chamber after lunch. Entering the decrepit army hut that passed for a laboratory, I heard the recording motor on the ionisation chamber wailing like an upset banshee. The motor was hard against its safety limit. I immediately suspected an electrical problem, turned the beast off and started to hunt for the problem.

Luckily, my research supervisor, Geoff Fenton, came in just then to perform his lunchtime check of other instruments in the lab. Seeing me head down, tail up in the electronics cabinet, he said, 'Ken, the counting rates of all the cosmic ray telescopes are very high. Have you been using an electric drill?' Electrical drills were notorious for interfering with all electronic equipment.

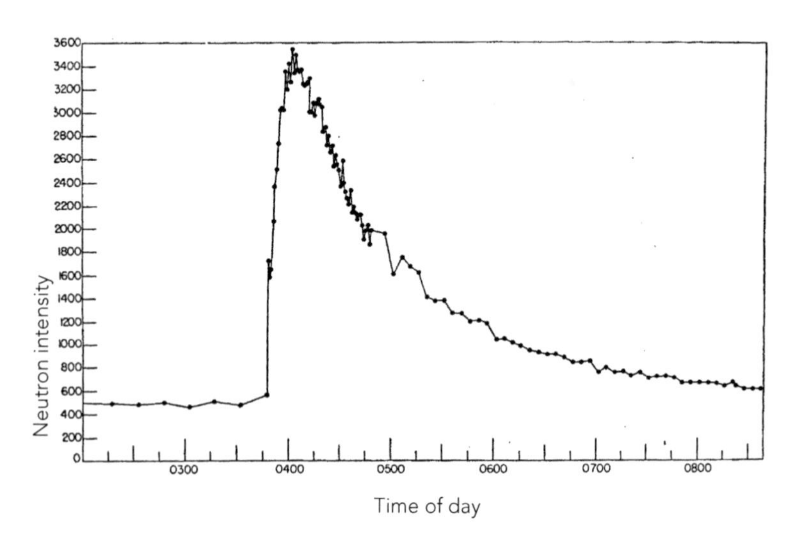

Figure 1: The very large cosmic ray burst of 23 February 1956 as observed in New Zealand. This alerted NASA to the radiation hazards that astronauts would face on the Moon.

I said a rude word, stood up quickly, hit my head on the overhanging electronic cabinet, said an even ruder word and shouted, 'Geoff, it's an *expletive deleted* solar flare'.

It was a beauty (Figure 1). The peak cosmic ray intensity was up to 40 times the normal value in some parts of the world. It was ten times bigger than anything the Sun has given us since. Examining the results obtained worldwide, I decided that something queer was going on, which the theory of the day could not explain. It made me wonder if the Earth's magnetic field was playing tricks on us. I decided some careful calculations of the effects of the Earth's magnetic field on the cosmic rays were desirable.

I mapped out the step-by-step calculation that would be required. In 1956 we did not have an electronic computer, but I was expecting to go to Antarctica or Papua New Guinea in 1957 and thought it might be a good project to do in my spare time using a hand calculator. When I estimated that, working eight hours per day, each calculation would take six months to compute with the fastest hand-computing machine available, I gave up on that little idea—for the moment, anyhow.

Working eight hours per day, each calculation would take six months to compute with the fastest hand-computing machine available …

Sunspots and solar flares soon became very important in my life in space research, so it might be useful to explain here what they are. A sunspot is a region of the Sun that is somewhat cooler than the rest, and therefore emits less light and appears darker than its surroundings (Plates 5a and 5b).

A sunspot is cooler because it is a region of very strong magnetic fields, up to 10,000 times stronger than the Earth's magnetic field. These fields extend over huge distances—up to 100,000 kilometres (60,000 miles)—and they wax and wane rapidly.

The magnetic field of a sunspot contains a prodigious amount of energy. Fairly frequently a hiccup occurs in these fields and, as a result, many hydrogen bombs' worth of energy can be released within seconds. This heats the upper layers of the Sun to very high temperatures, resulting in the emissions seen by Carrington, and the ejection of a prodigious quantity of solar material from the Sun at velocities of up to 2500 kilometres (1550 miles) per second.

Over the next few years, my calculations and space experiments allowed me to make important discoveries about these, and the properties of the so-called 'interplanetary space'.

‹ 8 ›

Lic Lic Skule Boi

The physics student in the 1950s encountered a number of risks in the pursuit of his or her studies: electrocution was the most common; poisoning by the chemicals we used with gay abandon was a possibility … But being a PhD student in Papua New Guinea meant that I had the added possibilities of being killed by a bold pilot or eaten by a puk puk or a kanaka.

It had always been my ambition to take my cosmic ray equipment to Antarctica. In the 1950s the conditions and activities there were still very much in the mould of those experienced by the epic explorers of the earlier decades—Scott, Amundsen and Mawson.

A small ship that would roll on a calm pond took you there in summer across 3000 kilometres (1900 miles) of seas known as the roaring forties. You could not leave until the ship visited again the following summer. A scientist had to take everything needed to repair the equipment, no matter what happened to it. It was a real challenge.

Somehow the scientific fates mixed up my wishes. I was given the job of establishing a cosmic ray observatory in Papua New Guinea. The instrument I had built for the Mawson research station, on the Antarctic continent, was taken there by a colleague from the University of Tasmania.

Installing my instruments in both Lae and Mawson would contribute to the International Geophysical Year, outlined in the previous chapter. My job was to build a neutron monitor and to take it and another cosmic ray instrument to Papua New Guinea. Just where in Papua New Guinea wasn't clear; I was told to 'find a good place'.

Some six months later my equipment was ready to go. It was packed into about 30 large boxes and about a hundred very small boxes. The small boxes were only about 500 x 250 x 100 millimetres (20 x 10 x 4 in) in size, but they weighed 50 kilograms (110 lbs) each. They were full of lead. Later they would cause me quite a bit of excitement.

Flying to Papua New Guinea was an experience in itself. The aeroplanes used were rather tired Douglas DC-6s. They flew at about 300 kilometres per hour (190 mph), and it took over three hours to go from Melbourne to Sydney. Then we started north, landing every three hours or so at Brisbane, Townsville, Cairns and Port Moresby, finally arriving at Lae on the northern coast of Papua New Guinea a day after leaving Sydney.

The air conditioning was primitive; the plane was noisy and uncomfortable. However, many of the passengers didn't seem to notice this. At first they were too busy having a boisterous party, and later they were in too great a state of intoxication to notice.

I didn't realise it then, but that flight north was one of the great institutions of those days, when Papua New Guinea was a territory of Australia, and the Australian population consisted, more or less in equal measure, of planters, policemen and missionaries. The former two categories seemed to regard life in the tropics as a great game, and parties were a very important part of that game.

And so we landed in Lae. A rather short grass runway, a cluster of old corrugated-tin aeroplane hangers and several wrecked Japanese ships in the harbour were all I could see.

Arrangements had been made for the Australian Department of Civil Aviation (DCA) to 'look after me' while I was in Papua New Guinea. No one really knew what that meant, nor did they really care—it was typical of the very informal arrangements that were the norm in the Territory.

To start with, I was taken to the DCA 'dongas', the living quarters where I would stay for the princely sum of $3 per week. Soon my 'wash boi', named Rar-wally, turned up. I didn't ask for him; it was just assumed by everyone—Rar-wally, the management of the dongas and the other inhabitants—that I would simply have to have one (Plate1).

Then an interesting problem developed. Being the tropics, and being in the tradition of the British Raj and all that, it was the done thing that I would be referred to as Masta Ken.

Now Ken is a relatively uncommon name and I had never before, nor since, had the problem of there being more than one Ken in circulation. Of the twelve inhabitants of the dongas, there were three Kens. The accepted procedure was then to use the surname—Masta Smith or Masta Jones. Us Kens weren't to be accommodated that easily, though.

One had the surname Bation, but he absolutely refused to be called Masta Bation. For me, it was assumed that the native Papua New Guineans would have trouble pronouncing my surname, so my official name became Masta Mac.

One had the surname Bation, but he absolutely refused to be called Masta Bation.

I soon learned that I had another name. It seems that someone had explained that I had come from 'big pella schule bilong Australia'. The creation of a university in Papua New Guinea was still many years in the future and this was the only way to describe the concept at the time in pidgin English. Somehow the idea of an over six-foot-tall Masta, who towered over most of the Europeans and all of the Papua New Guineans, yet was still at school caught the imagination of the locals. To all and sundry I was known as 'Lic Lic Schule Boi' (little school boy).

The Department of Civil Aviation gave me the use of a truck, driver and four 'bush kanakas' (tribesmen) and told me where I could find my pile of boxes in a 'warehouse' on the wharf. The warehouse was the first surprise—it looked like a large chicken house with many birds enjoying the scenery from the top of my pile of equipment. I tried unsuccessfully to use my very limited pidgin English to tell the kanakas to load my boxes onto the truck. Luckily, two teenage Australian schoolboys—home for the school holidays—had attached themselves to my party and provided vital translation skills.

Thirty very big boxes and 100 very little ones! There are no prizes for guessing that the kanakas all went to the little boxes. The first tried to pick one up, then all hell broke loose. They all ran back several metres and stood shouting and gesticulating at my little boxes. They were very clearly of the view that there was something decidedly indecent about little boxes being so heavy. Not only indecent, but unhealthy too, as far as I could tell. It took a lot of coaxing before I had all of my boxes where I wanted them.

After several weeks the equipment was busily counting cosmic rays and there was little left for me to do other than wait for it to break down. Which it didn't—by then we Tasmanians had learnt how to build reliable equipment—and acute boredom set in.

Not for long, though. I talked to my friends in Civil Aviation and found that they had 'an arrangement' with the local airlines, which allowed them to fly anywhere in the Territory as a 'supernumerary' on the aeroplanes of Mandated Airlines. As a supernumerary, you sat in the jump seat behind the pilots and occasionally did useful things, such as telling the passengers to do up their seat belts.

The passengers were primarily tribal Papua New Guineans, who would be travelling as indentured labourers to a plantation in another part of the Territory. To complete my side of the bargain I would bellow out at the top of my voice, 'Yu kissim seat belts quick time, eh—maski buggerim up'.

The aeroplanes used were some of the most wonderful of all time—the Douglas DC-3. Most of them were left over from World War II. They had no passenger seats, as we know them—just two hard benches down each wall of the cabin. Before the passengers boarded, the space down the middle of the cabin was stacked almost to the roof with cargo. Cement, food, car tyres, pigs and the like were all piled in place and covered with a heavy, tightly stretched net. The passengers then clambered on as best they could and found somewhere to put their legs among the cargo.

Some of the mountains of Papua New Guinea are very high; even some of the mountain passes are higher than the operating ceiling of the DC3 and most of the other aircraft in use in Papua New Guinea at the time. In the tradition of the bush pilot, these pilots had a solution to this minor problem. I experienced it on one trip flying from Manus Island (an island north of the main island of New Guinea) to Lae. There was a simple route, around the coast past Finschhafen, but there was another way, through the mountains, which was—to the pilots at least—more 'interesting'. I was sitting in the jump seat and soon realised that we were flying up a narrow valley and that the mountains on either side were very high. Looking at the altitude meter, I saw that the DC-3 had nothing more to give in the altitude department.

I then saw that we were heading for a mountain pass, even though the pass was about a thousand feet above our maximum altitude. Luckily, I had heard of this practice and nonchalantly continued to talk to the pilots. We kept flying up the valley, moving ever closer to the jungle canopy. When it seemed that we were about to plough into the jungle, there was a sudden sense of acceleration upwards, as the plane was pushed up by a strong wind blowing up over the pass. I asked the pilots what they would do if the wind happened to be blowing the wrong way. They just rolled their eyes. At such times one remembered the saying about Papua New Guinean pilots, 'there are old pilots and there are bold pilots, but there aren't any old bold pilots'.

I then saw that we were heading for a mountain pass, even though the pass was about a thousand feet above our maximum altitude.

There were other risks in Papua New Guinea in those days. The 'puk puk' (the crocodile) was perhaps the most notable. Big puk puks were occasionally seen at the end of the Lae airstrip not far from where my equipment was installed. They frequently snatched dogs for dinner and it was clear that they would enrich their diets with a PhD student if they were given the chance. They were sometimes seen in the river near the dongas where I lived. We paid them great respect.

They were not the only risk to life and limb. At that time, about half the map of PNG had 'UNCONTROLLED' printed on it in bold letters. This was still tribal country, and the centuries-old cycles of tribal warfare continued unabated.

I once flew to Mount Hagen, close to the border with Irian Jaya (West Papua), and learned of the consequences first hand. Just the previous year, a police party had walked into the middle of a tribal fight. The two white policemen leading the party were eaten by the kanakas.

The physics student in the 1950s encountered a number of risks in the pursuit of his or her studies: electrocution was the most common, poisoning by the chemicals we used with gay abandon was a possibility and occasionally there were quite pedestrian risks such as being crushed by a tonne or two of lead.

Being a PhD student in Papua New Guinea meant that I had the added possibilities of being killed by a bold pilot or eaten by a puk puk or a kanaka. It made for an interesting time and I enjoyed it very much. It demonstrated, I guess, that I was not risk averse.

After four months my cosmic ray equipment was still behaving itself and I had run out of interesting things to do on the side, so I handed the equipment over to the technicians of the Department of Civil Aviation and flew back to Tasmania.

Little did I realise then that my stay in Papua New Guinea would lead to a life in space research. Two years later I was invited to work in America with Bruno Rossi, a world leader in cosmic ray research. Bruno later told me that his decision to invite me to work at the Massachusetts Institute of Technology was influenced by his view that anyone who could operate cosmic ray equipment in the equatorial jungles must be a competent experimentalist. The work I did while working with Bruno led to my first satellite experiment in 1963.

‹ 9 ›

Sputnik I

The idea that the 'artificial moon' would keep circling the Earth was totally new and strange … The ABC radio station … explained that the satellite was like the tea that doesn't fall out of the billy as you whirl it around and around with the lid off.

It was the spring of 1957. Driving home to Hobart from a day's walking in the central Tasmanian mountains, I heard a news flash on the radio. The Soviet Union had launched an artificial moon, *Sputnik 1*.

In 1957 most people had no idea of what that meant. There was a conventional view that 'what goes up must come down when the noise stops' and, if it were 100 odd miles up, it would come down with a hell of a bang once the rockets stopped firing. The idea that the 'artificial moon' would keep circling the Earth was totally new and strange.

The Australian Broadcasting Association's radio stations went to great trouble to explain that there was no danger and there were erudite discussions, which left everyone totally confused. It was explained that the satellite was like the tea that doesn't fall out of the billy as you whirl it around and around with the lid off.

As I recall it, the commercial radio stations took the other approach and beat up the fact that a well-known Sydney department store had taken out an insurance policy against their customers being hit by a falling satellite while in the store.

A well-known Sydney department store had taken out an insurance policy against their customers being hit by a falling satellite while in the store.

When I arrived home my mother said, 'the professor has phoned for you several times this evening. He said it was very urgent'.

I called him. 'Ah, McCracken, about this artificial satellite. Harvard University has sent me a telegram. They think it may be a fake. They wonder if we can provide verification.'

Why ask Tasmania? Well, Tasmania was one of the few places in what was called 'the free world' where the satellite would go overhead an hour or so after sunset. The satellite would still be brightly lit by the Sun, and would appear as a bright moving star. If it were there, we would see it.

And why me? The Soviets had announced that the satellite was broadcasting on radio frequencies of 20 and 40 megahertz. The professor knew that I was a radio amateur. My radio transmissions had ruined a number of his experiments into the electrical currents that flow in plants. I suspect he thought, 'at last he can be some use'.

I turned my radio equipment on. As I expected, the radio amateurs of Hobart had not wasted any time. They had already 'tuned in' to *Sputnik*, and I was quickly told that, 'Bill VK7YY saw it going overhead at 9 pm'.

I called Bill. He was the senior telegraphist at the Overseas Telecommunications Commission and therefore a very unflappable, methodical person. 'What luck!' I thought, 'Just the right person to provide hard factual data'. He could read very fast morse code and was frequently involved when emergencies occurred in the rough seas around Tasmania.

'Yes, Ken', he said, 'I certainly saw it about 11.03 Universal Time. I was listening to it on 20 megahertz and it became very loud. Then the carrier frequency started to drop quite quickly'.

'That's right, Bill. That's the Doppler shift due to its speed'. I thought, 'That will convince the doubting Thomases at Harvard'.

'I ran outside the shack' (amateur-speak for wherever they had their equipment) Bill continued, 'and there it was—about 45 degrees from the horizon, to the south. It was very bright and moving quite quickly. Funny though, Ken, its brightness seemed to fluctuate every ten seconds or so.'

'That's okay, Bill. That was probably because the carrier rocket was rolling end over end.'

Continuing, Bill said, 'It passed overhead and kept going to the north. I last saw it about 10 degrees above the horizon'.

'What an excellent observation!' I thought. 'Bill, how long would you say it was from the time you first saw it to when you lost it on the northern horizon?'

'Well, Ken, it was going very quickly, some 20 to 25 seconds I guess.'

'Bill, are you certain. Could it have been two or three minutes?' I asked, knowing that, from what he had told me, it would have been more than five minutes.

'No, Ken, absolutely not.' At that stage of my career I had not experienced how excitement or danger can totally destroy the ability of even very objective people, such as Bill, to make accurate observations.

I drafted a telegram for the Prof to send to Harvard. It mentioned the Doppler shift, the fluctuating brightness, the direction it was going and the time it was seen. It also stated that the observer's estimate of the time taken to cross the sky was clearly wrong—suggesting that this was due to excitement.

The Americans had hoped that it had been a hoax. They had confidently planned to launch their own artificial moon. They were absolutely certain that they would be first into space. So they named it, with great pride, *Project Vanguard*. Bill's observations told them that they would be second into space behind the Soviets. This was not good news at all.

We Australians quickly renamed the American satellite *Project Guardvan*. For those who were born after we stopped using them, the guardvan was a special carriage at the end of a train. It was always the last to arrive at the destination. When I moved to America two years later, reference to our Australian name for *Project Vanguard* would initially lead to comments such as: 'How quaint, what does that mean in your language? Is it another one of those quaint medieval words that you English use—like fortnight and motor car?' When told that a guard van in Australia was what the Americans called the caboose, it suddenly seemed neither quaint nor humorous to them. I learned to curb my humour, somewhat.

I digress. Back to Hobart on the night of 4 October 1957. About midnight I 'tuned in' *Sputnik 1* myself. Its signal consisted of short beeps, which I thought would be conveying its measurements to

Earth. Then I realised that there was very fast morse code there as well. I concentrated and wrote down what was being sent, letter by letter. It said, 'Hi there. I am the man in the moon', repeated over and over. It was clear that the idiot fringe had found *Sputnik* already. It would have been very easy to tune a transmitter to the frequency of the satellite and program a morse code machine to send the message over and over. Who knows, it may have been a mob of university students.

There was very fast morse code there as well ... It said, 'Hi there. I am the man in the moon', repeated over and over.

The next night we decided to have a *Sputnik* observing party at the university. About 50 people turned up for the first time we would all be able to see *Sputnik*. It was too low on the horizon, and while we could hear it going by, we could not see it. The next 'overpass' would be about 100 minutes later, and a fair amount of alcoholic refreshments had been brought 'just in case'. As the time of 'overpass' approached, the throng climbed onto the roof of the old army hut that housed the physics department, at considerable risk to life and limb. I was listening on the radio. I called out, 'I can hear the Doppler shift. It must be visible by now'.

We knew that *Sputnik* would be coming from the south. However, one of our number was perversely looking to the north. (He later went on to be a high-ranking official in the Australian defence services. Perhaps looking in the wrong direction was one of the job selection criteria.) Anyway, he shouted out, 'There it is, low down in the northern sky'.

Everyone turned around, thinking that we had somehow missed it as it went overhead. For the next minute or so, there were heated and somewhat inebriated arguments as to where it was, whether it was moving and so on.

Meanwhile I was becoming very excited. 'The radio signal is very loud. It must be almost overhead.' It was. With everyone looking the wrong way, *Sputnik* had climbed up in the sky and already passed overhead. So much for science and scientists, it seemed to be saying.

> *With everyone looking the wrong way,* Sputnik *had climbed up in the sky and already passed overhead.*

Looking back, those 24 hours were an uncanny prediction of the years ahead: nothing ever happened as planned. There was an ever-present level of chaos and yet, by improvisation and by grabbing any chance that went by, a successful outcome was usually the result. Come to think of it, that is a good description of any newly emerging scientific endeavour.

‹ 10 ›
Jim van Allen

The newly discovered radiation belt was immediately called the van Allen belt. Much was made of the fact that the American satellites had made a major scientific discovery, which had eluded the Soviet Union with its much bigger satellites.

In the 1950s the cover of *Time* magazine provided a quick snapshot of the march of events in America and some other parts of the world. It would display photographs of presidents, politicians, philanthropists, film stars, spies and the odd gangster. There would even be the occasional clergyman or engineer, but very seldom scientists.

Therefore, the 4 May 1959 issue of *Time* magazine was particularly unusual. It featured a physicist, Professor James van Allen, from the University of Iowa in the corn belt of America. The article inside included smaller images of a recently graduated assistant professor, Ernie Ray, and two graduate students, Carl McIllwain and George Ludwig. Over the previous year they had discovered the radiation belts that encircle the Earth and had begun to repair the American self-confidence that had been sorely damaged by the Soviet Union in 1957.

The launch of *Sputnik 1* had been a devastating shock to American pride. At first the Americans could not believe that it weighed 84 kilograms (185 lbs). The *Vanguard* satellite they were building weighed only 9.8 kilograms (22 lbs). It was seriously suggested that the Soviet press release had put the decimal point in the wrong place. The American rockets were smaller than those of the Soviets and did not have enough oomph to orbit anything heavier than 9.8 kilograms. The Americans were appalled that the Soviets were able to orbit such a large satellite, which also meant that they would be able to launch very big atom bombs in their direction—if push came to shove.

Americans could not believe that it weighed 84 kilograms; the Vanguard *satellite they were building weighed only 9.8 kilograms. It was seriously suggested that the Soviet press release had put the decimal point in the wrong place.*

To further encourage the Americans to do better, only one month later the Soviets launched *Sputnik 2*, weighing around 500 kilograms (1100 lbs). In addition, it had on board the cosmonaut dog, Laika. It was a clear sign that there would be a human in space soon—and it would be a Soviet citizen.

The American press was outraged when *Sputnik 1* was launched ahead of their own satellite, and even more with the launch of *Sputnik 2*. The government's official response was, 'Let's not panic. Our *Vanguard* satellite will soon be in orbit and it will be a really clever satellite, not like the great hunk of metal launched by the Ruskie'.

The satellite survived, fell into the scrub adjacent to the launching pad and started to transmit as if it were in orbit. The press called it 'Flopnik'.

In private, Dr Wernher von Braun and his military sponsors exerted great pressure to be allowed to have a go as well as the navy, who were building *Vanguard*. On the night of launch of *Sputnik 1*, von Braun had said they could have a satellite in orbit in 60 days. After a good night's sleep, it became 90 days. Even that was an incredibly short time.

The reason: von Braun and two other research groups had always had an unofficial Plan B, which was almost ready to go. The Jet Propulsion Laboratory in California and Jim van Allen in Iowa had planned a small, simple satellite several years previously to be launched on the *Redstone* rocket, the American version of von Braun's *V2* rocket.

Some say that it was the German heritage, others that the science seemed too simple—whatever the reason, they never received the go-ahead. Now they were given a half-hearted nod to get ready, but not to launch yet. *Vanguard* was still seen as the real answer to *Sputnik*.

It didn't work out like that at all. The *Vanguard* rocket blew up on the launch pad in December 1957. The satellite survived, fell into the scrub adjacent to the launching pad and started to transmit as if it were in orbit. The press called it 'Flopnik'. A second one also failed. Worldwide, the cartoonists had a marvellous time.

Plan B went into effect. when the five kilogram (11 lb) *Explorer 1* roared into orbit in January 1958. The only scientific instrument on the satellite was a simple Geiger counter, which measured the electrons and protons passing through it. No great surprises were expected; some regarded it as trivial science. The data radioed back to Earth by the satellite was also less than encouraging at first.

There were no recording systems on the satellite, so data was only received when the satellite was over one of the few American satellite receiving stations scattered around the world. Over some, the Geiger counter was happily reporting that it was picking up radiation. Over others, however, it said that it wasn't picking up any radiation at all. That seemed most unlikely and the concern was that the Geiger counter was faulty. Understandably, this was not reported to the press.

Another satellite was rapidly cobbled together. The rocket blasted off without problems, but the satellite was never heard from again. Finally *Explorer 3* was successfully launched in March 1958. By then one of the graduate students, George Ludwig, had built a small tape recorder, which was flown in the satellite to fill in the big gaps between data receiving stations. Each time the satellite was in view, the tracking station would send a radio signal, which caused the tape recorder to send its recorded data back to Earth.

Looking at the data, the Iowans began to understand. Above some parts of the Earth, the Geiger counter would sedately count the cosmic radiation arriving at Earth from out in the galaxy. Moving along its orbit, the counting rate would steadily, then rapidly increase to quite enormous values, indicating extremely intense radiation. Then suddenly, the counting rate would drop to zero. Some time later, it would suddenly start to count again at a frenetic pace, then decrease back down to the low cosmic ray rate. Orbit by orbit, this pattern was repeated.

The penny dropped. The Geiger counter was tricking them, big time. When exposed to extremely high radiation, it simply stopped working. It was known that Geiger counters did this, but it never occurred to anyone that there would be such intense radiation up there. The theoretician of the team, Ernie Ray, used the results to demonstrate that the radiation was trapped in the magnetic field of the Earth.

It was a scientist's dream. A simple experiment had discovered a totally unexpected result. They had discovered that there was an intense belt of radiation encircling the Earth.

The American press went into overdrive. The newly discovered radiation belt was immediately called the Van Allen belt. Much was made of the fact that the American satellites had made a major scientific discovery, which had eluded the Soviet Union with its much bigger satellites.

The Iowans built more satellites. They included Geiger counters to better map the regions of high radiation intensity. A magnetometer was included to measure the Earth's magnetic field out in space. Over the next several years, Iowa City, a small university town in the middle of the American farmlands became the centre of the world's scientific exploration of space.

So how did an obscure professor in a minor regional university end up rescuing the American scientific reputation?

Jim van Allen, known to friends as 'Happy' because of his permanent smile, had grown up and been educated in Iowa. He started work as a nuclear physicist at the Carnegie Institution of Washington in 1939. The Carnegie Institution had a long history of careful and innovative research into the magnetic field of the Earth and other associated phenomena. In particular, it was operating the first worldwide network of cosmic ray recorders, which was beginning to discover the things I would study in Hobart fifteen years later. Jim became quite interested in this 'geophysical' research, and became acquainted with some of the world leaders in it.

Jim's research at the Carnegie Institution was cut short by World War II. He was assigned to assist with the development of proximity fuses for use in anti-aircraft shells, which made the shells go off even if they were going to miss their target. Perhaps that was how he became attracted to things that went bang and moved at great speed. Whatever the reason, Jim was given the job of developing the *Aerobee* rocket after the war in 1946. The *Aerobee* was a 'sounding rocket', which would fly to a height of 200 kilometres (125 miles) before falling back to Earth. 'Sounding' in the sense of a mariner taking measurements with a lead line on sailing ships, known as depth sounding. In this case the instruments on the rocket measured the temperature, composition, density and other properties of the atmosphere. The *Aerobee* and similar sounding rockets were the mainstay of atmospheric and space research in the days when it was called 'upper atmosphere research'.

One night in 1950 a number of the world's top geophysicists had dinner at Jim's home. During the evening they conceived the idea that there should be a concerted, international investigation of all geophysical phenomena in 1957–58, the 75th and 25th anniversaries of the International Polar Years of 1882–83 and 1932–33 (collaborative efforts to research the polar regions). As mentioned in chapter seven, 1958 was predicted to coincide with a period or high solar activity, promising many important discoveries might be made. The mechanics of international and national scientific politics were set in motion, leading to the International Geophysical Year (IGY) of 1957–58. That dinner led to the satellite programs of America and the Soviet Union and ultimately, to space research as we know it today.

Jim was appointed as the head of the department of physics at the University of Iowa in 1951. He and his students started to investigate the properties of the upper atmosphere using balloons—a cheap, do-it-yourself approach.

To go even higher, he developed the 'rockoon'. In this, a small, military rocket—left over from the war—would be attached to a balloon. The rocket would be ignited once the balloon had risen above most of the atmosphere. Without the frictional effects of the atmosphere, the little rocket would go much higher than if it had been launched from the surface of Earth. Flying these rockoons from the Arctic, the Iowans saw some unexpected electrons, which they associated with the aurora borealis, the 'northern lights'. Later they would realise that these may have been electrons leaking out of the radiation belts.

The research group in Iowa continues to be one of the leaders in space research to this day. Jim van Allen continued to provide instruments for pioneering missions into deep space for many years. His instruments detected and mapped the extremely intense radiation belts around Jupiter and Saturn. His instruments were on the *Pioneer* spacecraft that have gone to the far extremes of the solar system (see Chapter 33). His students and research associates have gone on to provide other key measurements of the space around us.

Jim remained active in space research until he died in 2006, aged 91. I exchanged emails with him as recently as late 2005; his memory was sharp, and his interest in space and research matters unabated.

The radiation belts are called the van Allen belts in honour of their discoverer. To me, however, the name means much more than that. To me, the name honours a man who was a key player in the early stages of the Space Age, and who then 'led from the front' in scientific research for almost 50 years. Not for him a prestigious job in Washington, nor in a large aerospace company. He was a fine example of a good scientist who had no ambition other than to be an even better scientist, to impart to his students and others the fascination, practical know-how and rigorous thought that characterised this remarkable man.

‹ 11 ›

Cupid in the Space Age

A drive to the laboratory took you through this fine forest, with views stretching to the islands at the mouth of the River Derwent. So it was quite natural that I asked Gillian … 'Would you like to drive up Mount Wellington with me to my observatory to collect the photographic recordings …?'

In the years following the launch of *Sputnik 1*, first the Soviet Union, then America launched a number of satellites that were easily visible for the hour or so after dusk. Sometimes the rocket that launched the satellite would go into orbit as well; they were large and were even visible from cities, despite the scattered light from streetlights.

The nightly procession of satellites across the sky fascinated Australians for a number of years. Having overcome their initial concerns that the satellites would suddenly decide to fall out of the sky, parents would take their children out in the backyard to 'see the satellites going over'. Today, 50 years later, many people have vivid childhood memories of this activity.

The Space Age also provided an entirely new form of assistance to the progress of romance. 'Going out to look at the satellites' became a welcome addition to the reasons that a couple might go walking together. Some of the satellites were quite faint, so scientific rigour decreed that you should walk as far away from streetlights as possible.

'Going out to look at the satellites' became a welcome addition to the reasons that a couple might go walking together.

My research also gave me another avenue for romance. I had established a cosmic ray observatory in a small hut on the slopes of Mount Wellington, just outside of Hobart. Mount Wellington is a relatively high, rugged mountain with a dense eucalyptus forest that extended to an altitude just above my cosmic ray laboratory.

A drive to the laboratory took you through this fine forest, with views stretching to the islands at the mouth of the River Derwent. So it was quite natural that I asked Gillian Filby, the young lady who would later become my wife, 'Would you like to drive up Mount Wellington with me to my observatory to collect the photographic recordings for the last week of operation?' Alas, this almost became my undoing.

We drove up to a point some hundred metres from the cosmic ray hut. We walked up to it, and I checked that all the bits were blinking and clicking like they should. I removed the film from the camera, which automatically recorded the cosmic rays every hour, then walked down to my car. To my horror, I realised that I had locked the car keys in the car. No matter what I did, I couldn't find a way in.

The only solution was to walk down the mountain to the nearest tramline, 10 kilometres (6 miles) away. Luckily, I knew all the tracks on the mountain very well and we were able to find the fastest route. This wasn't so smart either, since Gillian, not knowing we were about to go bushwalking down tracks at an angle of 30 degrees from the horizontal, was wearing high-heeled shoes and a very tight skirt. This was not a promising way to commence dating a young lady, but it seems that somehow it was all forgiven.

Months passed, then I saw a small four-line advertisement in the weekend edition of the Hobart *Mercury*. It sought applications from scientists to be considered for research fellowships in America. It gave no further details, except that a doctorate (PhD) was necessary.

I didn't yet have a PhD, and there was every probability that it could be two or more years before I did or that I would never achieve one. However, I have never been one to let small details interfere with my actions, and I sent off an application.

Four months later I received a letter offering me a position with Professor Bruno Rossi at the Massachusetts Institute of Technology (MIT), in Boston. This generated an enormous crisis of confidence.

Bruno Rossi was part of the explosion of talent in nuclear physics in Italy during the 1930s. He was the first graduate student of Enrico Fermi, one of the leading theoreticians of his day. Bruno invented an electronic circuit that could decide whether two electrical pulses occurred within a fraction of a millionth of a second of one another. Without it, cosmic ray and nuclear research would have been impossible. While very young, he had shown that an extremely pushy recipient of the Nobel Prize was wrong in his assertion that cosmic rays were gamma rays. He had written the definitive book on cosmic ray physics, every page of which reinforced my inferiority complex. On top of that, MIT was one of the two most prestigious engineering universities in America. My reaction was, 'I'm going to make a complete idiot of myself'. I seriously considered not accepting. But 'Nothing ventured, nothing gained', as they say.

A large number of forms arrived in which I had to state that I was not intending to overthrow the American Government nor was I going to engage in immoral activities and other things that were out of vogue in America at the time.

These did not present a problem, but one of the other forms did. It required that I state whether I was married and whether my wife would accompany me. This question was obviously of great interest to the bean counters, because my stipend would rise from $10 to $11 per day. (This is not a misprint. The American Government deemed that a wife could be supported on $1 per day. For children there would be an additional $1 per day). Needless to say, a tax rate of 30 per cent would apply.

The American Government deemed that a wife could be supported on $1 per day.

The simple answer was that I was not married. However, our courtship (as we called it then) was progressing in a satisfactory manner, and there seemed to be distinct possibilities, but then, maybe not. After all physicists can be odd at times, and any sane lady would think twice about putting up with such strangeness.

So I decided on a 'conditional proposal'. After discussing the pros and cons with my beloved, I moved to the next stage in those more formal days. I visited her father to ask for his consent, but with the explicit caveat that we were not quite certain. The key issue was that I would have his permission to include her on the forms I would send to America as my 'bride to be', and I would also need quite a lot of personal information about his family and that of my potential mother-in-law just in case a marriage eventuated!

Her father and mother were well aware of the blossoming romance. When Gillian told her father that I wanted to have a small chat with him, they guessed part of the reason. In the event, he was even more nervous than I. He repeatedly tried to divert me from my purpose. He asked questions about things like astronomy, the moon and satellites. When I eventually reached the point, he agreed without any questions. He never did tell me what he thought about the conditional nature of the proposal.

Months passed. Gillian and I went walking one Sunday night after church 'to watch the satellites going over'. As we walked, the full moon rose, big, bright and quite beautiful in the unpolluted Tasmanian sky. I started to waffle on about space research and how satellites would be sent to the Moon in the next year or so; about how scientists would be sent there not long after, to perform experiments and to set up astronomical observatories among other things. These would be just like the scientific bases in Antarctica, where some of my cosmic ray recorders were already clicking away with gay abandon ...

Gillian stopped dead in her tracks. She knew of my thwarted ambition to go to Antarctica. She also knew of my irresistible urge to climb all of the mountains in Tasmania, the harder, the better. She could see, writ large, where this species of waffle might lead. 'Promise me,' she said, 'that you will not go to the Moon, or anywhere else in space'. It was clear to me that a half-baked Tasmanian scientist was highly unlikely to ever get within a bull's roar of that possibility, and I readily made the promise. Little did I suspect that I would be crawling around inside a prototype *Apollo* spaceship within five years.

And so Gillian and I sailed off to America in the Orient line ship, the *Oronsay*, in July 1959. We had only been married three months. Together, we sailed off to a life in space research.

‹ 12 ›

Satellites—up close

I saw a large object with black paper and aluminium cooking foil wrapped rather roughly around it, with lots of sticky tape and a number of electronic cards lying around it, as if they had been thrown there from across the room.

Arriving in Boston in 1959, I walked up the imposing steps and through the huge Corinthian columns that formed the entrance to the Massachusetts Institute of Technology (MIT). I walked down the kilometre-long 'infinite corridor' which connected the densely-packed buildings on the immense campus. I had known that MIT was big, but this first ten minutes told me how much I had underestimated its size.

Bruno Rossi was a world leader in cosmic ray and, now in 1959, space research. After graduating in Italy, he had rapidly made his mark as an excellent experimental physicist with a very good understanding of theory. He built a cosmic ray experiment, which used the Earth's magnetic field to identify the nature of the cosmic radiation. He took it to Eritrea in much the same manner as I took my equipment to Papua New Guinea two decades later.

I was luckier than he. I was allowed to put my equipment in a tin shed. All he was given was a tent.

The Mussolini regime had cut off all support for scientists such as Bruno who were from the 'wrong' ethnic background. Bruno then went to the University of Chicago in America to continue his cosmic ray research.

With the initiation of the Manhattan Project to build the atomic bomb, Bruno and his family moved to the research laboratories established at Los Alamos, a remote spot in the mountains of New Mexico. There he was responsible for the sensors that measured the radioactivity which was a central part of the project. His final job was one that remained secret for 40 years—the measurement of how fast the bomb 'went off'.

Those days in the mountains of New Mexico had an enormous impact upon all those who were there. It was a hotbed of talent—many of the leading physicists who had revolutionised nuclear physics over the previous decades were there, together with many of the brightest new minds. This interaction, and the need to move quickly, meant that the younger people matured greatly in a very short time.

When the war ended, the best universities went headhunting at Los Alamos. Bruno and several of his co-workers were recruited by MIT. He then gathered around him a group of incredibly bright people. Like him, they were equally good at experimental and theoretical physics. With a ready source of money for research, they commenced the study of the cosmic radiation in great detail. They looked for cosmic rays from the centre of our galaxy or from the Sun, and tried to understand what was then known as 'interplanetary space'.

Then the Soviet Union launched *Sputnik 1*. As I mentioned in Chapter 10, the Americans were appalled. The ignominious failure of the *Vanguard* satellite made it worse. Finally, almost six months after *Sputnik 1*, *Explorer 3* led to the discovery of the van Allen belts. This was a great moment. However, the small size of the satellite was still regarded as a national disgrace.

American officialdom went into overdrive. The press wailed about the Soviets winning the Space Race, about the inferiority of American technology and education, and so on. Congress became very upset and threw great wads of money at the problem. More money for education, for research, for bigger and better rockets and, in particular, for space research.

So was born the National Aeronautics and Space Administration—NASA. This was to be a civilian organisation, and the universities were encouraged to think of ways to use the rockets and money it would have. The people at MIT were quick to respond.

Many forms of radiation—gamma and X-rays, ultraviolet and infra-red radiation—cannot penetrate the Earth's atmosphere. A satellite flies above the atmosphere, so an instrument in a satellite would be able to see the radiation previously hidden from us. The MIT people designed the instruments, and received piles of money to test them in the laboratory and in high-flying balloons. Rockets were ordered by NASA to fly these instruments into space.

It was into this hive of activity that I was thrown in August 1959.

Bruno was quite vague about what research I should do. There was a large cosmic ray detector on the laboratory roof and he suggested that I might supervise the students working on that. Clearly he thought I might like a larger challenge so he suggested 'look around and see if there is something that interests you'.

The first laboratory I went into was the home of the 'MIT Gamma Ray Satellite'. I looked around and saw the usual clutter of electronic bits and pieces, tools and the physicist's construction material of last resort—sticky tape. 'Where is it?' I asked, expecting to see a bright shiny, gold-plated masterpiece of the engineering art.

I looked around and saw the usual clutter of electronic bits and pieces, tools and the physicist's construction material of last resort—sticky tape.

'There', I was told—and I saw a large object with black paper and aluminium cooking foil wrapped rather roughly around it, with lots of sticky tape and a number of electronic cards lying around it, as if they had been thrown there from across the room. Around them was a jumble of coloured wires connecting the electronic cards; what in physicist speak was called a 'rat's nest'.

'But that looks like the sort of stuff I have been building in Tasmania', I thought. Suddenly I realised that satellites were not always fancy or flash. From that moment the idea started to grow that I would like to design and build a satellite instrument of my own. Just one day previously that had been an impossible dream—I had arrived in America assuming that my experimental and engineering skills would never allow that—but now ...?

Across the corridor I entered the 'Plasma Probe Laboratory'. The plasma probe was to be flown halfway to the Moon to see how empty space really was. At that time there were many theories and few facts. The textbooks said that the space between the planets was almost empty. A common view was that there was less than one atom per cubic centimetre. Eugene Parker, however, a young theoretician from the University of Chicago, had recently proposed that that there was a 'solar wind' blowing away from the Sun at 450 kilometres (280 miles) per second. As a result there could be a hundred times more gas in space than previously thought. Gene Parker also predicted that there would be a very weird magnetic field extending from the Sun to the Earth, which was quite unlike anything previously predicted. Another theoretician argued for a much slower 'solar breeze' of 20 kilometres (12.5 miles) per second—less gas but another weird interplanetary magnetic field. MIT intended to find out the facts.

The plasma probe looked just like a small metal chamber pot or—for people who do not remember the days when the toilet was a little shack in the furthest extremity of the backyard—a metal saucepan without a handle, with several layers of metal mesh and wires across the mouth of the saucepan. It didn't look at all complicated. I knew that there would be some pretty ferocious problems due to the bright sunlight in space, but it was explained that some clever tricks with the voltages on the mesh and wires could overcome that problem. Once again I thought, 'Hey, this doesn't look that hard. Simple and clever'.

The plasma probe looked just like a small metal chamber pot ...

I decided not to join either of those projects and started to 'do my own thing', using a very large computer—large for 1959 that is—in the basement at MIT to calculate the effects of the Earth's magnetic field on cosmic rays coming from the Sun. The differences in the Australian and American educational systems meant that I was younger than most of the American PhD students and became a member of the 'student bunch' at lunch each day. These students were among the very best at MIT and in America. They were doing a great deal of the research and construction of the MIT space experiments, so I quickly learned firsthand what space research was really like on a day-to-day basis. It confirmed my impressions from that first day—that space research was little different from the work I had been doing in Australia.

At lunch, listening to these top-flight American students, I gradually came to learn an interesting fact, which began to counteract my pronounced inferiority complex. I began to realise that while the American students knew an immense amount of theoretical stuff, they often did not know how to use it. Furthermore, their skills in the laboratory were usually quite limited, and they were reluctant to try new things. Having been a virtual one-man band for five years, building and operating instruments in Papua New Guinea, on mountains and for the Antarctic, perhaps the only thing I was really good at was trying things out in the laboratory without asking anyone. 'Suck it and see' was the technical term for this approach. Sometimes this resulted in plumes of smoke, or a pile of worthless numbers, or a few days of wasted effort. However, you soon learned how to be a good 'experimenter'.

I also learned to ask 'dumb' questions. Scientific discussions at universities can come close to the limit of our knowledge—and to be truthful they frequently contain wishful or personal views that verge on pure bullshit. However, expressed properly, the purest of bullshit can sound like reasoned scientific fact. Nevertheless, there is a strong tendency for young, inexperienced scientists not to ask questions for fear of being thought to be dumb. Perhaps it was the example set by Bruno Rossi, but I learned to ask those simple questions or query the assumptions people were making. It was astounding how often it stopped people dead in their tracks.

Scientific discussions at universities can come close to the limit of our knowledge—and to be truthful they frequently contain wishful or personal views that verge on pure bullshit.

Over the next year my work with the computer in the basement at MIT was successful. The Sun was also very cooperative and launched several great bursts of radiation into space. Using my computer calculations I was able to show that the radiation reached Earth from the direction exactly predicted by the 'solar wind' concept. This was the first experimental demonstration of the validity of the theory that soon became accepted by everyone. I became famous, in a rather niche market.

The two MIT experiments were launched into space soon after. The plasma probe on *Explorer 10* was first to go. As planned it went about halfway to the Moon. For half of that distance it remained inside the Earth's magnetic field and gave low readings as expected. Then it passed out into 'interplanetary space', and it immediately registered a strong 'solar wind' blowing from the Sun in the vicinity of 450 kilometres (280 miles) per second. These field observations agreed with the predictions of Eugene Parker from the University of Chicago and with the results I had obtained from my analysis of radiation bursts from the Sun. Seventy hours after launch the batteries ran out (this was before the days of solar cells in space); the satellite died and crashed back to Earth.

However, during its short life it had totally changed our understanding of 'space', and MIT had achieved a double whammy. *Explorer 10* had measured the properties of the solar wind and the interplanetary magnetic field in a completely unambiguous manner, and I had shown that the magnetic field in space stretched all the way from the Earth to the Sun exactly as predicted by Eugene Parker. To a rather unconfident graduate of the smallest university in Australia the feeling was electric. I had participated in a major scientific breakthrough.

Then the gamma ray telescope was flown on *Explorer 11* a month later. It orbited the Earth for some months. It saw gamma rays coming from the Milky Way and led to an entirely new form of astronomy.

It was pioneering science. I was well and truly hooked. I started to design experiments that I thought should be done in space. However, first I needed to convince NASA to spend millions of dollars on my ideas. I was becoming well known in my field, though, and I thought, 'Perhaps, just perhaps, that will do the trick'.

A year later it did.

‹ 13 ›

My ticket to space

You told the computer how to make your calculation by punching holes in paper cards. Purists wrote the instructions in 'machine language', which consisted of a string of ones and zeros … Pragmatists, like me, used a 'programming language', which was really a pidgin English for computers.

I had always liked arithmetic and calculating things. Moving quickly past simple sums, I took to geometry, algebra and trigonometry like a duck to water. It was natural then, I suppose, that I took great interest in ways to calculate things quickly.

First there were logarithms. To multiply 2 and 3 you looked up a table and read that the logarithm of 2 was 0.30102, and of 3 was 0.47712. You added them up and made 0.77814 then looked that up in an anti-logarithm table which gave you 6. Fast it was not.

Then I found the joys of the slide rule. It was like a chunky ruler, with a groove down the centre, in which another thin ruler would slide. You lined numbers up, fiddled the centre ruler back and forth a bit and read off the answer. Multiplying 2 and 3 you would arrive at 6.01. Fast but not accurate.

Multiplying two and three you would arrive at 6.01.
Fast but not accurate.

Then I graduated to calculating machines. The first of these had a number of knobs, which you moved to indicate the first number to be multiplied. To multiply that number by 237 say, you would turn a handle clockwise seven times, push a button, turn the handle three times, push the button again and turn the handle twice. To divide, you turned the handle anticlockwise.

Then the high-technology machines arrived. An electric motor did all the turning of the handle three times and pushing of the buttons stuff. It was as noisy as hell, but you didn't have sore muscles at the end of the day.

As a consequence of the large cosmic ray burst in 1956, I had become seriously interested in computing on a large scale. As mentioned in the previous chapter, I wanted to calculate the path followed by electrically charged cosmic rays as they passed through the Earth's magnetic field. To do this the path would be broken up into thousands of small steps. The starting and ending points of each step would be calculated, one after the other.

I worked out the equations I would use to calculate each step, then estimated how long it would take to compute a single path using a desk calculating machine. I found that by working eight hours a day, it would take six months to achieve just one result. I needed hundreds. Clearly the turn the handle and push the button technology would not do.

Arriving at MIT in 1959 I found that there was an IBM 704 electronic computer in the basement—that is, in the *whole* basement. By today's standards it was brain dead, but in those days it was the fastest in the world. I redid my calculations to see how long it would take to compute a cosmic ray path. The answer: twenty minutes. Clearly a big improvement on six months.

I found that there was an IBM 704 electronic computer in the basement—that is, in the whole basement ... The instruction book ... gave me the cheerful news that it would break down roughly every ten minutes.

However, there was a problem. This was the era before transistors, integrated circuits and microprocessors. The computer was full of 'radio valves', which became very hot and failed with monotonous regularity.

The instruction book for the computer gave me the cheerful news that it would break down roughly every ten minutes. It suggested that the computer would not last the distance for even one of my twenty-minute calculations.

Nevertheless I decided to go ahead. Many hundreds of calculations later the computer had not broken down once—when I was using it.

In the 1950s the press referred to the electronic computer as the 'electronic brain'. The computers of the 1950s were not 'brainy'. They had an IQ of an unusually dopey Irish setter and were much less enjoyable to know.

The computers of the 1950s were not 'brainy'. They had an IQ of an unusually dopey Irish setter, and were much less enjoyable to know

You told the computer how to make your calculation by punching holes in paper cards. Purists wrote the instructions in 'machine language', which consisted of a string of ones and zeros that told the machine the steps it was to take. Pragmatists, like me, used a 'programming language', which was really a pidgin English for computers. It consisted of a very limited vocabulary of English words, which the computer could translate into its own machine language.

One hole in the wrong place, one missing full stop or bracket or a typographical error and the computer 'threw you off'. Since I might have the use of the computer for only a few minutes each day, being thrown off by a dumb computer was not appreciated.

Having spent a month or so learning the programming language, and punching the 300 cards that told the computer what to do, the next job was to 'debug' it. That is, to make sure that the answers were correct. This was not as easy as it sounds.

Since I had not spent six months calculating a correct answer, how was I to know what was correct? I could check for some types of errors by using the computer to follow a hypothetical cosmic ray from a point on Earth to a point in space. I would then 'turn the particle around' by changing its electrical charge and squirt it back towards the Earth. If the particle ended up back at exactly the same point on Earth where it had started, some types of errors were not present. I also discovered that, ten years previously, several Swedish scientists had used a laboratory model to study the motion of cosmic rays in the simplest approximation of the Earth's magnetic field (the dipole). These measurements allowed me to check for other simple errors.

The real worry was that I was using a complicated formula for the Earth's magnetic field, which had 48 different terms in it, with up to the sixth power in trigonometric functions. This gave enormous scope for something to go wrong. Firstly physicists always tend to leave factors of two and pi (3.1416) out of their equations by mistake—and I have a very bad dose of that disease. Then the trigonometric functions are alternatively plus and minus, and it is easy to screw that all up. The chances were very large that there would be an undetectable error. There were also no published results of the results to be expected from the simulation of the field I used, so I was well and truly on my own.

I devised some checks to test for the worst types of errors, but could do nothing else to improve my chances. However, science is often like that. First you check your assumptions. Then you do what you can to check your results, then, if you have confidence and nerve, you present the results at a scientific conference or two, but you are always thinking about two questions: 'Is there some other way to check the result?' and 'Will someone think of something that will prove that I am wrong?'

This time I was lucky enough to be able to answer the first question well. Let me explain: the Kerguelen Islands are a bleak group of islands somewhat to the north of Antarctica. They are on the 'great circle' route from Capetown to the eastern states of Australia. In the 18th and 19th centuries, many ships were wrecked as they sailed through the islands at night or in fog. The islands had their uses, however. Captain Cook visited them to obtain the Kerguelen cabbage, which he fed his crew to prevent scurvy. (These cabbages continued to grow there until some intellectual genius introduced the rabbit from Australia—they denuded the place.)

The Kerguelen Islands are part of the far-flung French nation, and the French maintain a research station there as part of their Antarctic research program. Among their instruments was a 'neutron monitor' similar to the ones I installed in Tasmania and Papua New Guinea. It was with this monitor, one day in 1960, that the scientific fates gave me the assistance I needed.

There was a great explosion on the Sun, and the cosmic ray recording instruments all over the world saw a huge burst of cosmic radiation. Comparing all the results, they were all in close agreement, all except those from Kerguelen Island.

The data from here was seriously weird. It was quite wrong for the latitude of Kerguelen; the results were what you would expect from an instrument 10 degrees further south in Antarctica. There was much scratching of the pencils of the theoretical physicists, but to no avail. It was even suggested that the French equipment was not working properly, and in those days of 'le grand Charles' (de Gaulle), this was taken as a terrible insult.

Now, it was known that the Earth's magnetic field misbehaves in the vicinity of South Africa, so I thought, 'Could the unusual magnetic field be the cause of the weird results obtained at Kerguelen?' I plugged the latitude and longitude of Kerguelen into my computer program. I instructed it to see what would happen for a particle energy that conventional theory said would never reach Kerguelen. The computer told me that, to the contrary, it would zip into Kerguelen with the greatest of ease.

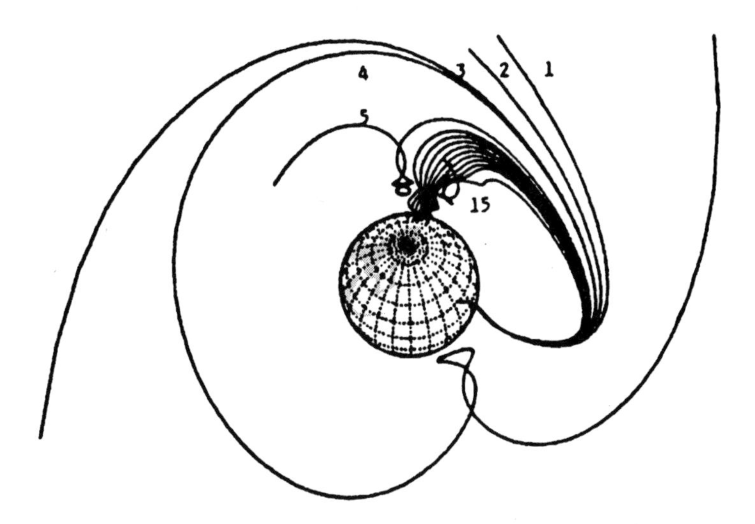

Figure 2. Illustrating the paths followed by cosmic rays of different energies before they reach Washington, DC. The highest energy particles (orbits 1 & 2) are relatively simple and the cosmic rays reach the Earth with little deflection. The low energy particles (4, 5, 15) follow very convoluted paths. For this case, the cosmic rays would reach Earth from many different directions in space.

This single calculation told me that my program was in partial agreement with the observations. I kept running the program, one calculation each night. After three weeks it was clear that the computer program was in excellent agreement with the French observations—and this agreement told me that I had the mathematics, and the computer program, right.

So I published the results of all my work using the computer program—four separate papers—in one issue of the world's most prestigious journal of geophysics, the *Journal of Geophysical Research*. I gave the program to anyone who wanted it, so they could run their own computations. However, the second question was still in my mind; would I receive a letter some day saying, 'I have found an error in your calculations'?

I received that letter three years later. It came from a distinguished theoretician in the top theoretical laboratory in space physics in America. He said I was wrong, and criticised me for 'errors in the magnetic force terms'. He gave no proof that I was wrong. He said that he had written the program correctly, and was now going to repeat the work I done.

I was just an experimental physicist with only three years of university tuition in theory and mathematics and had a severe inferiority complex when it came to 'real' theoretical physicists. I grew very worried. I wrote to him, outlining the tests I had done, and suggesting that he use his program to compute a number of test orbits, both for the simple dipole field and for the complicated simulation of the Earth's magnetic field that I had used. I mentioned the agreement between my calculations and the observations from Kerguelen Island.

He wrote back saying that tests were pointless because my equations were simply wrong.

I wrote back, asking this distinguished 'expert', 'Are the errors you are referring to in my coordinate acceleration terms?' This was an allusion to the complexities that occur when the mathematics are written in what is known as 'spherical coordinates'. I mentioned that the coordinate accelerations tended to be overlooked—as I had, in fact, done at first, and only woke up to the problem when my results were clearly crazy.

Silence. No acknowledgement. No reply. In fact, I never heard any more about that work. Nothing was published in scientific journals. Several years later I actually met him at a scientific meeting and reminded him of our exchange of letters. With bad grace he said, 'We made a mistake'.

Clearly that was the end of *that* subject. He walked off and I learned that experimentalists with little theoretical knowledge can, in fact, beat theoreticians at their own game.

I'll return to MIT … Once I had 'debugged' my program it was ready for use. As I left the institute each evening, I would punch the cards needed for five or six new calculations. The following morning they would be finished and I would think, 'Another three years of work completed'.

Quickly, however, the inevitable happened; MIT gave me a ration of two hours of computer time every three months. I scrounged more. Within weeks it was all gone. Luckily, I mentioned my problem to Frank McDonald, who was the leader of one of the scientific programs at the Goddard Space Flight Center near Washington, DC. His response was quick: 'Ken, we have three IBM 704s. Would you like me to get you the night shift on one of them for a week or two?'

Today the Goddard Space Flight Center is a densely packed conglomeration of laboratories extending over a large area … of Washington, DC. Then it was two lonely buildings standing in a large cow pasture, which had been churned to mud.

So, in the winter of 1969 my wife and I and our six-month-old daughter drove to Washington in our old heap of a car that

I had bought for $200—and had been robbed blind at that price. Luckily, we also found a new way to avoid the toll on the tollway—the car caught fire when we stopped to pay. The officials of the Massachusetts Turnpike Authority seemed to be more concerned about their nice shiny toll booths than in asking for our money.

Today the Goddard Space Flight Center is a densely packed conglomeration of laboratories extending over a large area on the outskirts of Washington, DC. Then it was two lonely buildings standing in a large cow pasture, which had been churned to mud. Reaching it at midnight was always an adventure.

Once in the computer room, I would settle down to the night's work. I would start the computer going with a card deck that specified the first few cosmic ray paths to be calculated. The theory was that I would then punch the next deck of input cards while the first computations were being made, and so on.

The theory quickly broke down. At first I would have only five minutes to spare after punching the new deck before the computer needed it, but quickly the computer began to catch up. After about an hour it would become a neck-and-neck race.

In retrospect, I had a puritan view of that computer. People of my generation were the children of Depression-era parents, and our earliest years were during World War II. If we didn't eat our meals, our parents would say, 'You mustn't waste food, think of the starving children in China', or 'Waste not, want not'. So I had this real phobia about wasting the time on the IBM 704, as if my mother were saying, 'Think of all the post-doctoral fellows who only have an IBM 1604', which was a *real* clunker of a computer.

I had this real phobia about wasting the time on the IBM 704, as if my mother were saying, 'Think of all the post-doctoral fellows who only have an IBM 1604', which was a real clunker of a computer.

When I went in on the second night the atmosphere was different. The two IBM 704s that I was not using had a rope around them, and signs on them saying 'No Entry'. There was quite a number of people there, looking at the printed output that was streaming out of the computer printers. There seemed to be a lot of talking on telephones.

'What gives?'

'Oh. Goddard is backing up Cape Canaveral for the launch of the monkey into space'.

So that was it. This was the trial of the *Mercury* spacecraft that would, some months later, take the first American astronaut, John Glenn, into orbit in 1962. The monkey was hooked up to all the medical monitoring equipment that would later be used on humans.

'Backing up' meant that the Goddard computers were performing the same calculations that were being made at Cape Canaveral, so that if the computers at the Cape were to break down, Goddard would be able to take over. All the calculations to decide if everything was working, to track the spacecraft around the Earth and to monitor the health of the monkey were being performed next to my computer. It was an interesting night.

By my fourth night I was feeling the strain. I was worn out with running the race against the computer. Then my brain did a strange thing. I suddenly doubted that the computer could be doing six months worth of calculating every few minutes. No amount of logical thought would make the doubt go away. It made me make errors in punching the cards, and the computer started 'throwing me off'.

I decided there was no alternative other than to perform a calculation that would prove to my tired mind that the computer was fair dinkum. I prepared an entirely new set of input cards for one test path I had computed many times when first assessing the program. It gave the right answers, and put my mind at rest. While it lasted, however, it was a very disquieting feeling.

At the end of the week we returned to Boston with a great pile of calculations. They kept me busy for years. I don't believe I ever used my computing program again. Peggy Shea from the University of New Hampshire asked me if she could use the program for her research. She and her husband have used it ever since and have become the world experts on the effects of the Earth's magnetic field on cosmic rays.

For me, the program had done what I wanted—and I didn't know it then, but these calculations had also opened the door to the whole of my scientific career. They allowed me to deduce a number of important scientific results about the Sun and interplanetary space. Those, in turn, earned me the opportunity to fly my instruments on seven NASA satellites. That, in turn again, led to my subsequent career in the CSIRO. It is sobering to think that probably none of these things would have happened for me if, in that first month at MIT, I had not embarked on what appeared to be a foolhardy enterprise.

‹ 14 ›

Magnetic expressways in space

The strange magnetic field would give us warning of the huge magnetic clouds that sometimes come rushing out from the Sun and play merry hell with radio communications, pipelines, electrical power distribution networks and other important parts of our modern society.

The scientific fates are often quite capricious. You might have a glimmer of a new idea, but find it impossible to prove. To better understand the matter, you will perform experiments that are designed to eliminate some of the possible explanations of what you have seen. With luck, perseverance and the cooperation of nature, you may then make an important scientific discovery. This is what happened in my first six years of investigations as a space scientist.

In Chapter 7 I described the unexpected results obtained from the great solar flare in February 1956. Those results suggested that the cosmic rays were coming from the wrong direction in space. However, to be certain that the magnetic field of the Earth was not playing tricks on us, I later performed hundreds of calculations, as described in the previous chapter. With them, I was able to determine where the cosmic rays that arrived at a detector on the surface of Earth had come from.

Looking at those calculations I saw a most interesting result. The cosmic rays that reached Hobart, or Washington, DC, for instance had come from many quite different directions in space outside the Earth's magnetic field. However, the situation was quite different for the cosmic rays reaching locations close to the North or South Poles. For a detector at Mawson, the first Australian base on the Antarctic continent, the cosmic rays had all come from a small bundle of directions out in space.

I realised that I would be able to map how cosmic radiation depended on direction out in space if I had the data from a reasonable number of such high latitude detectors, each looking at the radiation coming from its own individual part of the sky.

About 50 neutron monitors had been established worldwide during the International Geophysical Year. However, most of these were at universities or research laboratories in civilised parts of the world, and only one was close enough to the North Pole to be useful to me.

About ten instruments, however, had been established at research bases in the Arctic and Antarctic, and calculations showed that all of them would be ideal for my purpose. Almost immediately after I realised this, on 4 May 1960, the Sun lived up to its obligations, and produced a large solar flare, which showered the Earth with cosmic rays. The results were seriously weird; even the most casual examination of the data obtained worldwide said that the cosmic rays were not coming from the direction of the Sun.

The results were seriously weird; even the most casual examination of the data obtained worldwide, said that the cosmic rays were not coming from the direction of the Sun.

Using my calculations, and the data from the cosmic ray detectors in the Arctic and Antarctic, I saw that the cosmic rays from the solar flare had arrived at the Earth from a direction about 50 degrees away from the direction of the Sun. Furthermore, the great burst of radio waves from the solar flare told me that the cosmic rays had arrived at Earth only several minutes later than they would have arrived if travelling at the speed of light. Clearly, they had not dawdled on the way from the Sun. Something had guided them unerringly to Earth.

Considering these and other facts, I came to the conclusion that there was only one possible explanation: cosmic rays carry an electrical charge and therefore a magnetic field forces them to follow curved paths. Most of the time they will go round and round in circles and take a rather long time to go anywhere.

The only way the solar cosmic rays could reach Earth quickly, from one direction in space, was if there was a magnetic field stretching from the Sun to the Earth, which arrived near Earth from a direction 50 degrees away from the direction to the Sun. Figure 3 shows the diagram I published in 1962 to illustrate my conclusion.

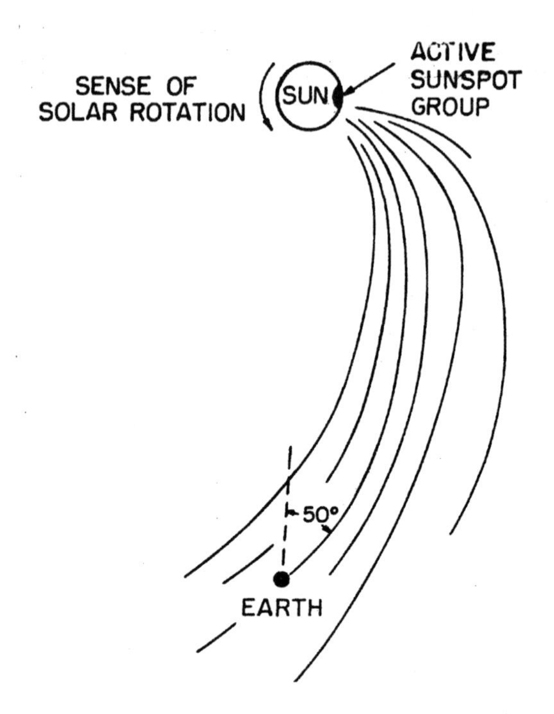

Figure 3: The diagram published in 1962, illustrating the author' conclusions regarding the nature of the magnetic field between the Sun and the Earth. Scientists call this the 'garden hose' magnetic field, because its shape resembles the water from a garden hose as the gardener swings the hose from left to right.

As I mentioned in Chapter 12 there was at that time a major controversy raging among the theoretical physicists who were studying the properties of 'interplanetary space'. It started when a German astronomer, pondering the strange behaviour of the tails of comets, had proposed that there was a 'wind' of gas blowing away from the Sun at high speed.

The most senior and highly respected space scientist in the world said this was all bunkum and that the strong gravitational field of the Sun would mean that there would be very little gas in space.

After careful study of the properties of the corona—what you could call the outer 'atmosphere' of the Sun—the University of Chicago's Eugene Parker came to the conclusion that the million-degree temperatures high above the Sun would cause gas to stream away at a velocity of about 450 kilometres (280 miles) per second. Such high velocities caused serious unrest among his fellow theoreticians, and other theories were advanced.

Another very reputable theoretician used complex mathematics to predict that there would be a solar breeze blowing at 20 kilometres (12.5 miles) per second. In those days before satellites had ventured far from Earth, it seemed that there was no way to tell which theory was right.

However, any gas in space exhibits a very strange property, which means that it carries magnetic fields along with it. That is, gas leaving the Sun stretches the magnetic fields from near the Sun far out into space. Looking at the three theories—no wind, solar breeze, and solar wind—they all gave beautifully different predictions regarding the magnetic field between the Sun and Earth.

If the Sun let out a great burst of cosmic radiation, the three theories predicted that the radiation would arrive at Earth in quite different manners:

- The no wind theory—predicted that the cosmic rays would appear to come immediately from the direction of the Sun.

- The solar breeze theory—predicted that the cosmic rays would take many hours to arrive here, and that they would come from a direction 90 degrees away from the Sun.
- Eugene Parker's solar wind theory—predicted that the cosmic rays would arrive quickly at Earth from a direction of about 45 degrees from the Sun.

My calculations, and the data from the solar flare in May 1960, provided incontrovertible evidence in support of the solar wind theory.

The Christian Science Monitor is a well-known newspaper in Boston and America. It takes a keen interest in the advances in science—in America and worldwide.

So, soon after I published my results about the interplanetary magnetic field, including the diagram in Figure 3, a reporter rang me to ask for an 'in-depth' interview. In describing my discovery, it became clear to me that I needed to find a simple way to explain how the magnetic fields 'told' the cosmic rays where to go.

Boston is an old city for America, and the city roads clearly show that they started life as the cow and horse paths through the original village. Driving through the curved, narrow streets of the city was a nightmare, so it is not surprising that one of the first by-pass, or 'loop', expressways in the world was built around Boston.

Named Route 128, it became famous as the home of a number of American high technology industries in the 1950 and 1960s, and as a pioneer of the by-pass roads that soon appeared all over America, and later worldwide.

Searching for a simple way to explain what the magnetic fields were doing to the solar cosmic rays, I said, 'It's just like the way you can get quickly from the northern to the southern suburbs of Boston by heading west on Route 128, which guides you quickly in a circular route around the chaos of the city'. The reporter liked this, and we further developed the analogy.

Some weeks later, a long article appeared in *The Christian Science Monitor* describing my research and using the 'expressway through chaos' concept to help everyone to understand the implications of my work. The American Moon program was in its infancy, and the article discussed the manner in which astronauts on the Moon might be in great danger if solar cosmic rays suddenly came rushing down the 'expressway'. Stated that way, everyone understood the implications.

The American Moon program was in its infancy, and the article discussed the manner in which astronauts on the Moon might be in great danger if solar cosmic rays suddenly came rushing down the 'expressway'.

I soon saw that the 'expressways in space' would have two other important effects upon us on Earth.

Firstly, they would 'focus' the cosmic rays travelling from the Sun. That is, while the radiation would start off going in all directions near the Sun, the magnetic field would quickly force them to be moving parallel to the interplanetary field when they reached Earth. They would all seem to be coming from the same direction in the sky.

Secondly, the field would cause then to 'bunch up', so they would all reach Earth at about the same time.

Then I saw that the strange magnetic field would give us warning of the huge magnetic clouds that sometimes come rushing out from the Sun and play merry hell with radio communications, pipelines, electrical power distribution networks and other important parts of our modern society. Ten, sometimes twelve, hours before they arrived they send a message through the strange magnetic field, which tells us that they are on the way.

In Chapter 24 I recount how my first discovery was used to protect the astronauts from danger while on the Moon. Both discoveries are used today to predict the 'solar weather' our satellites will see. To this day, the concept that the interplanetary magnetic fields act as an expressway for cosmic rays is still used.

My verification of Eugene Parker's concept of the solar wind and my insights into the nature of the interplanetary magnetic field were my first scientific discoveries of note. I was invited to present lectures at universities, to contribute to scientific encyclopedias and to provide contributions to some of the first books in the Space Age. My days at the Massachusetts Institute of Technology made me well known in space science. I was still only 29, an example of the fact that new fields of research are great places for young people to be.

‹ 15 ›

Skyhooks

There we were stumbling around in the dark preparing to launch a 100-metre long balloon. Until it was launched the balloon would be held down by an attachment to a very heavy truck. When released, the balloon would start rising very quickly and, some seconds later, snatch the experiment off the ground—and I mean snatch.

In the 1960s aspiring space scientists faced a difficult problem: how to gain the experience needed to build an experiment that could be flown in, and survive in space? Today, there are companies that have all the skills to build anything a scientist has ever convinced NASA to fly. In the 1960s you were largely on your own.

Testing an instrument destined for space was an even bigger problem. Most of the things we were interested in studying—for example, gamma and X-rays, short-wave ultraviolet light, cosmic rays, and the solar wind—could not reach the surface of Earth.

Some of them were deflected by the Earth's magnetic field, except near the North and South Poles. Nearly all of them were absorbed within the top two per cent of the atmosphere.

The balloons we were using were a far cry from the party balloons of our childhood. When fully inflated, they were about 100 metres in diameter and would lift a ton or more to an altitude of 32 kilometres and higher.

The only solution was to send your instrument to an altitude that was above at least 99 per cent of the atmosphere—that is to a height of 32 kilometres (20 miles) or higher. Aeroplanes could not reach those altitudes and there were only two other choices: the high altitude balloon or the 'sounding' rocket.

In the university community the balloon was the favoured method: it was cheaper, it gave a (fairly) gentle ride, it could carry quite heavy experiments—so you didn't have to miniaturise everything—and it could float for many hours providing much more data than the fleeting five minutes given by a sounding rocket. Most importantly, you were likely to receive your equipment back in one piece, allowing it to be flown again, after adjustments, if you were wrong the first time.

The balloons we were using were a far cry from the party balloons of our childhood. When fully inflated, they were about 100 metres (330 feet) in diameter and would lift a ton or more to an altitude of 32 kilometres (20 miles) and higher. They were made from very thin film, similar to that used to protect clothes after dry cleaning.

I built my first balloon payload in 1962, to try out ideas that I hoped I would fly in a satellite some time in the future. (Plate 3a)

Since it could be very hot or cold above the atmosphere, we tested the operation of the electronics and detection systems at temperatures between minus 30 degrees and plus 50 degrees Celsius (-22° to 122°F). To see the cosmic rays it was designed to detect, it had to be flown in the Arctic.

Each summer NASA and other American organisations organised a balloon flying expedition to Fort Churchill, on the shores of Hudson Bay in north-west Canada. This was a place of great history; one of the first trading posts of the Hudson Bay Trading Company was established there in the 18th century, and it still existed.

More to the point, America and Canada had established a large aerodrome here for the B52 bombers that would have been used to bomb the Soviet Union if the Cold War had suddenly warmed up a bit. The bombers were gone, replaced by intercontinental ballistic missiles, but the enormous concrete runways were absolutely ideal for flying the big 'skyhooks' we scientists needed.

My first flight was a disaster. The balloon must have had a leak in it, and it reached a height of only 20 kilometres (12.5 miles) before descending slowly back to earth—and that was it; I had been allocated only one balloon. After all, I was the novice who wasn't to be taken very seriously and, when the balloon was gone, well, tough. (Plate 3b)

A young scientist soon learns that networks are an important part of science. Within hours, and long before my payload had been returned to me from where it fell, 100 kilometres (62 miles) to the west, my friend Frank McDonald from the Goddard

Space Flight Center—who had given me the computing time described in the previous chapter—sought me out. He said he had a spare balloon that I could use. Two nights later our payload flew again, successfully this time, and we received the first proof that our ideas might work.

Another opportunity then appeared out of the blue. I received a letter from the US National Science Foundation, offering to provide the support to fly balloons from Hyderabad in southern India. I proposed that I should fly my payload, suitably adjusted for equatorial latitudes, to further test the ideas behind the satellite instrument we were by then building, at enormous expense, to protect the virility of the astronauts who would go to the Moon later in the decade.

The possibility of lost virility may have done it. Anyway I was told that they would provide the funds and balloons I needed—which meant hundreds of thousands of dollars.

The possibility of lost virility may have done it. Anyway I was told that they would provide the funds and balloons I needed …

Before I proceed, let me set the scene of those far-off days, when we knew stuff all about the Sun, and the space between it and Earth. Our Sun, a very ordinary star in the outer reaches of one of the 100 billion galaxies in the universe, has one minor claim to minor cosmological fame: it fluctuates in properties every eleven years (see Chapter 7).

The International Geophysical Year gave us a lot of new data and greater understanding of the Sun when it had lots of sunspots and solar flares. Our balloon safari to India was planned when we were approaching 'sunspot minimum'—scientists can make blandness into a virtue—so it was set for 1964, to be part of what the highest levels of scientific opinion decided would be called the Quiet Sun Year. 'Let's see what makes it tick,' the pundits said, 'when it isn't ticking!'

We modified our payload from Fort Churchill to measure the properties of any poor, unsuspecting cosmic ray that came our way. Then, just six weeks before we were due to leave for India, I heard exciting news. My mentor from MIT, Bruno Rossi, had flown a rocket to look for X-rays from the Moon, and had seen a great number streaming from somewhere near the centre of our galaxy. He had concluded that there was a star out there churning out great quantities of X-rays.

Now, the theoreticians had said that this was impossible. They said that there was no way that a star could churn out enough X-rays to be seen a hundred light-years away. Here was something new and totally unexpected. Was it, just possibly, the chance to get in on the ground floor of an entirely new branch of science? Every young scientist dreams of that.

I did some calculations. If there were enough X-rays to be seen in just five minutes—by the tiny little counter in the rocket with a diameter of only five centimetres (2 inches)—perhaps I could detect some of them from a high flying balloon. I could use a much bigger detector and peer out into space for hours, not minutes. The calculations were encouraging—it seemed that it might actually work.

I was not the only one to think this way. Another good friend from MIT, George Clark, was planning to do the same from Texas. Recognising my need for haste, George was good enough to provide me with the engineering drawings of the key component that would make his X-ray telescope scan the sky.

In less than six weeks I designed my X-ray telescope, built it and added it to the balloon payload that was nearing completion. Then into the packing crates it went, and off to India.

Our first balloon launching was set for 2 am. I chose this time so that the less bright part of the Milky Way would be rising in the east. As I saw it, my friends from MIT were looking at the brighter bit, and there was no point in repeating that.

So there we were stumbling around in the dark preparing to launch a 100-metre (330 feet) long balloon. Until it was launched the balloon would be held down by an attachment to a very heavy truck. When released, the balloon would start rising very quickly and, some seconds later, snatch the experiment off the ground—and I mean *snatch*.

Here one had to be very quick and careful. The experiment was on a small-wheeled cart, and, in the few seconds before the experiment shot into the sky, the cart had to be pushed quickly so as to be exactly under the balloon when it took up the payload. If not, the 800-kilogram (1750 lbs) payload would swing off the cart and crash into the ground, taking people and anything else in its way with it.

The big moment came. The scientists were shooed out of the way of the balloon launching crew—and nothing happened.

There was a bit of incoherent shouting; something to do with Fred, a young member of the launching crew. After several minutes all was ready again. The balloon was let go. It shot up into the sky. The little cart with my experiment on it was rapidly pushed to be under the rapidly ascending balloon.

There was a loud shout, 'Fred, GET OFF'. Fred had three fingers caught in the cables around my experiment and was rapidly being hauled towards the stratosphere.

Fred was not a happy lad, for a number of reasons. Just as the balloon was ready to go the first time, he experienced an excruciating attack of Delhi Belly. Too late, he ran behind a tree. Having abandoned some of his clothing there, he returned to launch the balloon. It was reported later that 'he was walking funny'.

So as he was pushing the little cart, his mind was not entirely on the business at hand. It is fair to say, however, that his mind became much more focussed as he felt himself heading for the stratosphere. When he was some four metres (13 ft) in the air, he shook himself free and tumbled to the ground.

... his mind became much more focussed as he felt himself heading for the stratosphere. When he was some four metres in the air, he shook himself free and tumbled to the ground.

As he did there was a loud explosion. The launch crew boss always had a little radio control box that would allow him to abort the launch in an emergency. By pushing a button, he could cause an explosive charge to destroy the balloon. This he had done, but the wind now dragged my experiment across the ground for several kilometres until it became entangled in a tree in Hyderabad's botanical gardens.

The next day my technician and I retrieved my mangled experiment. An enterprising Texan, my technician was convinced he could fix anything, anywhere. He headed off to the local Indian bazaar, hoping to find 'one-inch diameter 6061 aluminium tube'.

Indian bazaars are marvellous places. You can find bits for 100-year-old sewing machines; bottles of every conceivable shape and size; pots and pans beyond comprehension; but no 6061 aluminium tube. Indian traders are always keen to help, particularly when you have American dollars. So a group of traders gathered around trying to understand what was needed, in the hope of a big sale. After long descriptions of 6061 aircraft alloy tubing, one trader said, 'I have just what you want, come into my shop'.

So it was that our high-tech experiment—funded at great expense by the US space program—took off several nights later rebuilt from bamboo. It worked very well and we discovered the second high-energy X-ray star known to mankind.

So it was that our high-tech experiment—funded at great expense by the US space program—took off several nights later rebuilt from bamboo.

My Texan technician never admitted to the fact that he had used bamboo to build an experiment for the US space program.

‹ 16 ›

Spudnik

An experiment had been designed to find out whether plants would be confused in the absence of gravity. It was proposed that a potato would be flown to see which way it would sprout without gravity to tell it which way was up.

In NASA-speak it was called an 'AO', an 'announcement of opportunity'. It read rather like an advertisement in the lengthy matrimonial columns of an Indian newspaper: 'lonely temperamental satellite wishes to meet patient, broad-minded and tolerant experimenters'.

The year was 1963. NASA was still learning how to be a space agency. Up until then these AOs had been a mixed bag, to put it politely. Often the project would be cancelled after years of hard work. Frequently the rocket would blow up or the satellite would crash into the jungles of Africa. Sometimes the satellite would work well for a few days then, when everyone was congratulating themselves on being some of the lucky few who had actually managed to make something work in space, the satellite would have a hernia or something and it would never be heard of again.

Frequently the rocket would blow up or the satellite would crash into the jungles of Africa. Sometimes the satellite would work well for a few days then … have a hernia or something and never be heard of again.

Consequently AOs were regarded with some suspicion and this one particularly so—it was offering 'a ride', as we said in the trade, on a satellite that would travel from Earth to the vicinity of Venus. While this was ideal for my purpose—to measure the properties of the cosmic radiation far from the influence of the Earth's magnetic field—there were many things that could go wrong on a journey of 100 million miles.

There were other catches. The total weight of all of the experiments was to be just seven kilograms (15 lb). A total of only 4.5 watts of power would be available for all of the experiments.

For the greater part of its voyage the satellite would also send the scientific data back to the Earth at the mind-blowing rate of one byte per second. (For the computer nerds, that it not a misprint; I mean one byte per second. For normal people, this is like transmitting a two-digit number, say 43, every second, which is now regarded as an excruciatingly slow rate at which to transmit data). Oh yes, and the weight, power and data capacity were to be shared more or less equally between about five experiments.

I did manage to design an instrument that would measure the characteristics of the cosmic radiation, but the electronics were hopelessly overweight. It used transistors and they, together with the resistors and capacitors needed in the circuits, would weigh more than the allowance for all the experiments.

Some weeks previously I had hired an electronic engineer to assist my team. He said, 'Why don't you use integrated circuits?'

My reply: 'Integrated *whats*?'

I had a sinking feeling in the pit of my stomach that I was not going to cope with all this technological change. I had learned all my electronics using the old technology—vacuum tubes. Just two years previously I had taken the plunge and converted my thinking to transistors.

Now this guy was telling me that the transistor had been superseded already and that one of these integrated whatsits, weighing five grams (0.18 oz), could do what 200 grams (7 oz) of transistor electronics would do.

It was very depressing, but I carefully redid my calculations on the basis of using integrated whatsits. The tantalising result was that I could now build the whole instrument within a total weight of 2.1 kilograms (4.6 lb).

We put in a 'response to the AO' and explained that we would do many clever things. We said the instrument would weigh 2.1 kilograms, use 1.5 watts of electrical power and would need to send data back to earth at a rate of 15 bytes per minute.

When NASA received all the 'responses' a committee of experts decided who would get a ride.

Being early in the space era, there were relatively few people foolhardy enough to risk their sanity by responding to this one. However, there was one rather unusual response. An experiment had been designed to find out whether plants would be confused in the absence of gravity. It was proposed that a potato would be flown to see which way it would sprout without gravity to tell it which way was up.

An alert journalist heard of the proposed potato experiment. He wrote an article titled: 'NASA to fly *Spudnik*'. NASA did not think this was at all funny, but some of the more irreverent space scientists did. The name stuck for quite a while. Alas however, the potato experiment was not given a ride.

An alert journalist heard of the proposed potato experiment. He wrote an article titled: 'NASA to fly Spudnik*'.*

After the expert committee had pontificated for several months, a list of six experiments was announced for flight on the satellite. Three of the experiments had 'first class' tickets and were guaranteed to fly. Three had 'second class' tickets and would fly if the rocket could be given some extra oomph. My instrument had the highest priority second class ticket.

We learned that we were expected to build five identical instruments: one for practice, one to shake the hell out of to see what would break, two for flight and one as a spare (Plate 4a).

In those early days of space, we always assumed the worst and were pleased if that was all that went wrong. The satellite might become very hot or cold, so we had to test that everything would work in temperatures from minus 20 degrees to plus 70 degrees Celsius (-5° to 160° F). It would have a very rough ride on the rocket, so the instruments were subjected to a 'shake, rattle and roll' machine, which would simulate the intense vibration it would experience. To be on the safe side, the machine was set to give a 50 per cent rougher ride than was expected in real life. Then it would be tested for a long time in vacuum, to see whether this would destroy it.

All this made a lot of sense. However, our patience became rather strained when we started to deal with the 'quality assurance' characters. In theory they too were a great idea, they would make certain we used reliable parts and good construction practices.

Then we read the rules, which stated, roughly, that, ' You may use a component in an instrument for space flight provided that it has been flown in space before'.

We told the quality assurers what we would be using and were immediately told, 'You cannot use integrated whatsits or photomultipliers (a type of light amplifier). 'They're not on our list,' they said.

We argued, 'But these things are well-proven technology. Someone has to be the first to fly them'.

'No exceptions,' they said.

'But we said that we would be using both when we proposed, and we were chosen on that basis,' we said.

However, whatever we argued, these humourless characters told us, '*No* exceptions. If it's not on our list, you cannot use it'.

This was my first experience of bloody-minded, high-power bureaucracy and I began to develop my method of handling it: don't fight it, go round it.

I'll fill in a little of the background. While the world thought NASA was a single entity, it was far from that. It consisted of a headquarters, which everyone fought, and about six different 'centres' in different parts of the country, which fought to the death between one another.

Spudnik was being managed by the Ames Research Center, near San Francisco. It was a military-style organisation and did everything 'by the book'. Near Washington there was the Goddard Space Flight Center, which was run much more in the mould of a university engineering department, with a willingness to find creative ways around a problem. I checked the situation with Goddard, whose response was, 'Yes, photomultipliers and integrated circuits have been on the NASA list of approved components for the last nine months'. With glee they suggested that Ames was simply being tardy in changing their list to reflect the official NASA list.

So I sent a nice, polite telegram to Ames saying something like, 'Washington tells me that the components you say I cannot use have been on the approved NASA list for the past nine months. Hasn't the pony express reached you yet?'

As I mentioned, quality control people are quite humourless. They didn't think my telegram was the least bit funny. They quickly sensed the role of their mortal enemies at Goddard. With extremely poor grace we were told to 'get on with it'.

‹ 17 ›

There's many a slip …

We never found out what really happened … Our best guess was that they wanted our photomultipliers for a more important, or more persuasive, customer, and that they had sent us the standard unit hoping we wouldn't notice.

On 3 September 1963 we trooped into a small meeting room at the Ames Research Center of NASA, which was located in a technological enclave that later became known as Silicon Valley, California. There were about three people from each of the six teams who had first or second class tickets on *Pioneers 6* and 7, aka *Spudnik 1* and *2*. Most of the scientists knew one another and recognised that this was a rather elite group. We introduced the various engineers who were members of our individual research teams.

Then the spacecraft project manager led his team in. I thought the line would never end. Soon there were many more spacecraft people than experimenters. The thought dawned, 'We are about to ride a monster'. There was an assistant project manager, a chief scientist and umpteen engineers—communications, data processing, electrical, structural, power supply, rocket engineers and many more. Then there were the camp followers: the quality control people, the financial people and, of course, the public relations (PR) people.

Then there were the contractors' people. For, dear reader, NASA seldom built or designed anything themselves; this was usually done by companies well known for their work for the military or the communications industry. There were the people from the company that had the contract to build the spacecraft. The communications contractor was represented, and so on.

We quickly learned what the first American astronaut, John Glenn, was referring to when asked what he was thinking about while the countdown was nearing completion before hurling his Mercury capsule into space. He said, 'I was thinking that the two million parts of the rocket and spacecraft around me had been built by the lowest bidder'.

'I was thinking that the two million parts of the rocket and spacecraft around me had been built by the lowest bidder'.

Each of the NASA team and each of the contractors then started to tell us what we had to do for them. Little was said about what they would do for us. We were told that our satellite and its instruments had to be designed, built and tested to within an inch of its life, then launched on its way to Venus just two years in the future.

Well, we knew that. However, all of the experimenters were shocked out of their wits when we were told that our instruments had to be designed, built, tested and delivered to NASA within twelve months. We all said, 'That's impossible'.

We were all told, 'That's the way it's going to be'.

Then one of the scientific experimenters said, 'What are your design rules regarding magnetic cleanliness?' This was new to me. 'What does he mean by magnetic cleanliness?' I wondered. We were soon to find out, at considerable risk to our sanity and to the budget for the project.

To explain: in 1963 we had no good measurements of the magnetic field that pervades the solar system. The theoreticians had predicted that it would be small—about one millionth of the magnetic field here at the surface of the Earth. They said, however, that it might be ten times greater or smaller than this figure. So two of the big questions of solar physics were how big it really was and how much it would change, day to day and year to year. The answers would tell us a great deal about what made the Sun tick, and make the discoverer famous.

This was not an original thought. Having missed the chance to be the first nation to launch an artificial moon, America decided that it would be the first nation to launch a satellite that would escape from the gravitational pull of the Earth and go into 'deep space'. So several years earlier America had quickly built a spacecraft, put it on top of a very big rocket, and launched it into space, far, far beyond the Earth and the Moon. There were a number of experiments aboard that were to give the answers to the big questions of solar physics. A magnetometer was included to measure the solar system's magnetic field.

Pioneer 1 (as this craft was called) did not last long. While it was designed to send its measurements back to Earth from a distance of 10 million kilometres, it had a hernia, or some such and disappeared without trace about two million kilometres (1¼ million miles) from Earth. Before it did so, however, it provided magnetic results that were very perplexing.

Perplexing, that is, until some smart scientist was able to show that it was simply measuring its own magnetic field. For, while only a few things like iron and nickel seem to have their own magnetic properties, a great number of common materials 'soak up' the Earth's magnetic field and become weakly magnetised. Therefore, some of the materials on *Pioneer 1* took that magnetic field into space with them. Furthermore, as we all learned in high school, electric currents have a magnetic field around them.

In the haste to be the first into 'deep' space, the designers of the satellite had overlooked this. At great expense, *Pioneer 1* was measuring its own magnetic field and largely ignoring the weaker, but much more interesting, magnetic field streaming out from the Sun.

As we all learned in high school, electric currents have a magnetic field around them. In the haste to be the first into 'deep' space, the designers of the satellite had overlooked this.

This all helps to explain the question about magnetic cleanliness. The experimenter from the east coast of America had no intention, whatsoever, of repeating the mistakes of his arch rival from the west coast. He was determined, very determined, that this spacecraft was going to be squeaky clean, insofar as magnetic fields were concerned. At his insistence, very strict rules were set as to the materials we could use in building our experiments.

The rules for magnetic cleanliness made our job very difficult. We soon found that magnetic materials were used in the strangest of places. They were used for the wires going into transistors. Our photomultipliers (light amplifiers), which had already caused a severe chill in our relationships with NASA, were found to be full of magnetic wires. Seemingly innocuous things such as electrical wires were also found to have an iron wire down the middle, which threw the magnetic detector off the scale.

The photomultipliers were a source of great trouble. Not only were they 'dirty' in a magnetic sense, but conventional wisdom then said that they were susceptible to damage by the Earth's radiation belts and other penetrating radiation. We therefore requested that our photomultipliers should be constructed without the offending magnetic materials and out of light-amplifying materials that would be less sensitive to 'radiation damage'.

The price was enormous. The time to manufacture them was 90 days, which was 25 per cent of the whole time we had to build our satellite instrument. However, we were plunging headlong into the unknown, backed by the might of the US Treasury, so only the best would do.

Ninety days passed and the photomultipliers did not arrive. After 100, then 110 days I was in a flat panic. We sent telegrams;

we phoned and, eventually, 30 days behind schedule, we received them. Then we had to test that they were still efficient as light amplifiers after the changes we had requested. A colleague tested them immediately and soon reported triumphantly that we had not lost any efficiency. They were as good as if they had been made from the radiation sensitive material.

Murphy's well-known law states that: 'Anything that can go wrong will'. Murphy said nothing whatsoever about things going four times better than expected so I immediately smelt a rat. Perhaps, I thought, the photomultipliers were as good as if made from the radiation sensitive material, because they had been made from it.

I called the manufacturer, one of the best high technology companies in America and explained that the photomultipliers behaved exactly as if they had been manufactured from the radiation sensitive material. The sales engineer tried to tell me that my measurements were wrong. I convinced him otherwise and asked him to check his production records. He told me to, 'Call back tomorrow'. When I did, he said that he couldn't find them.

The NASA contract stated that careful records had to be kept of all the manufacturing processes. I explained that NASA would be very interested to hear that his records were missing. He found them. Then he couldn't read the writing. We circled round and round as I repeatedly asked, 'Did you, or did you not use the wrong material?' For 30 minutes he avoided giving an answer. Finally, with very bad grace, he admitted what we had suspected. They had supplied us with standard photomultipliers, while charging the exorbitant price quoted to provide the special units we had specified.

With that out of the way, I asked him what his company proposed to do to rectify the situation. The response was: 'There is nothing we can do. We could never make them in time for you'.

Suffice to say, I was not pleased. I sent a long telegram to the president of the company, explaining what had happened, stating that there had been repeated attempts to fob me off and informing him that I was now going to inform NASA about the whole sorry tale.

About an hour later, the administrative vice-president at our research centre was heard screaming to my boss about, 'his irresponsible people sending abusive and intimidating telegrams to the president of Corporation *X*'. He was even more upset that the said president was now on the phone wanting to speak to the person who had sent the telegram. Our vice-president made it very clear that something very terrible should happen to me.

So, I went to the vice-president's office and spoke to the president of Corporation *X* on a conference telephone. He immediately apologised and said they would replace the photomultipliers they had supplied with ones made to the correct specifications. I pointed out what his own engineer had said about it taking too long. His response was: 'Would five days be soon enough?'

After pointing out that it had taken them 120 days to make the wrong thing, I asked, 'Do you really expect me to believe you can now do a special order in five days?'

He said, 'We will do it.' They arrived five days later and turned out to be excellent photomultipliers, which showed no radiation damage whatsoever after many years in orbit.

We never found out what really happened. At no time did anyone tell us that standard units had been sent to us by accident.

Our best guess was that they wanted our photomultipliers for a more important, or more persuasive, customer and that they had sent us the standard unit hoping we wouldn't notice. When we caught them out, we surmised that they then sent us the units that had been made for us in the first place.

I learned a major lesson: know your facts ... and push hard. This has stood me in good stead over the years.

About a year before our first spacecraft was to be launched at the end of 1965, I received a letter from US Department of Immigration saying something like, 'Since you came to America on a visa that was valid for a maximum of two years and. since you are still here five years later, we believe it is time you left'.

This was the era when politicians the world over became concerned about the 'brain drain'. Since all the brains were draining to America, their politicians would occasionally tell their Department of Immigration to round up the usual suspects.

America had no intention of ceasing to attract the best scientists and engineers—this was the height of the Cold War and these people were the shock troops—but the right thing needed to be seen to be done, so I received my letter.

The letter caused some serious consternation on the part of the institution at which I was working. My contracts from NASA and other American funding agencies were supplying a substantial portion of all the institution's income and they didn't want to lose that.

Enquiries were made 'at a high level' and the institution was told that there would be no need for me to go, provided I obtained a letter from an appropriate authority in Australia saying that my home country didn't mind if I stayed down the drain, in a manner of speaking.

This seemed nice and simple, but who would give me such a letter? No Australian government authority would. I had no connection with any Australian university, so that was not a possibility. I wracked my brain and remembered that there was something called the Australian Academy of Science, in Canberra. I didn't know what they did, but the title sounded like it might impress the Americans, so I wrote a letter to them explaining my problem.

'This is to certify that it will be to the benefit of Australia if Dr KG McCracken remains overseas.' It didn't say 'for ever', but the tone of the letter gave that impression.

I didn't know it then, but the Academy was the most senior scientific organisation in Australia. Theirs was a select membership. As I had spent very little of my research life in Australia, they would know absolutely nothing about me.

Nevertheless, they understood the problem. Their executive secretary also apparently had a wry sense of humour. Thus it was that I received a letter addressed 'To whom it may concern', which read: 'This is to certify that it will be to the benefit of Australia if Dr KG McCracken remains overseas.' It didn't say 'for ever', but the tone of the letter gave that impression.

My American institution was delighted and the letter was sent to US Immigration. However, they didn't let me off the hook … yet.

I was informed that I would need to leave America, apply for a new visa and, if my case were good enough, I would be allowed back in.

After enough pain had been extracted, I was allowed to return to America and our first satellite roared into orbit a year later.

‹ 18 ›

An Australian family in Camelot

To my family and me, as Tasmanians, the whole of America seemed somewhat like Camelot. The high standard of living, the availability of good clothing in myriad sizes, the invigorating discussions with our peers, seemingly unlimited support for my research and the availability of a wide range of world-standard newspapers made for a heady mixture.

My adventures in space science during our first four years in America were amazing enough for a graduate of the smallest university in Australia. Our family life was no less remarkable. Gillian and I now look back at those years with a combination of awe and horror. 'Awe' because of the many life-changing experiences we had and the understanding of ourselves and the world that we gained so rapidly. 'Horror' because of the state of the Cold War at that time and our proximity to 'high quality targets'.

Within weeks of our arrival at MIT, an American U-2 spy-plane was shot down almost in the centre of the Soviet Union.

The American pilot, Gary Powers, survived and was shown on worldwide television as incontrovertible proof that America was spying on the Soviet rocket and nuclear activities. America said that it was a research aeroplane, which had been making meteorological measurements, but no one believed it. There was much sabre rattling and we checked our bank account to see if we could afford air fares home if things got too hot.

Some months later a group of Cuban exiles, with the support of the US government, staged the Bay of Pigs invasion of Cuba. By this time we had also learned that a large proportion of America's secret military research was conducted on Route 128, a ring road around Boston. This meant that our apartment and the Massachusetts Institute of Technology were at the bullseye of one of the highest-priority nuclear targets in America. We checked the bank balance again.

Three years after we arrived, we were moving from Boston to Dallas when the Cuban Missile Crisis erupted. The Soviet Union was setting up a missile-launching base in Cuba and a number of ships carrying more missiles were known to be approaching Cuba. America reacted strongly and armed hostilities seemed likely. Dallas is fairly close to Cuba and it seemed quite likely that it would be an early target if someone made a mistake. Luckily, the Soviets blinked, turned their ships around and the crisis was averted.

About that time Australia was making its first attempt to win the Americas Cup from the New York Yacht Club with the yacht *Gretel.* The television news gave roughly equal weight to the Cuban crisis and the battle for the Americas Cup. In some parts of the country it seemed that the possible loss of the Americas Cup was the greater concern.

The television news gave roughly equal weight to the Cuban crisis and the battle for the Americas Cup.

Our first year in America coincided with the preliminaries of the presidential election of 1960. We had inherited a television from a cousin who had been at Harvard the previous year, and we watched the Democratic and Republican conventions greatly amazed that the most powerful nation in the world seemed to conduct its electoral process in such a childish manner.

Then there were the Kennedy–Nixon debates. Gillian watched the televised debates from the United Nations. We were having a crash course in American and international politics. City politics also featured; we were in Boston when the Mayor was gaoled for some financial peccadillo. That in itself was not unusual in America; what was unusual was that he then ran for re-election while still in gaol and won by a handsome margin.

The presidential candidacy of John F Kennedy attracted strong passions. To many, the image of a young, vigorous aspirant was very exciting. To others the possibility that Kennedy would be the first Catholic president of America was deeply unsettling. Kennedy's stylish and pregnant wife Jackie Kennedy and their small daughter Caroline were very much the model American family. The newspapers and magazines couldn't stop writing about them.

The Kennedys were from Massachusetts and were known to be frequently in Boston. It became clear to us after a while that Gillian, who was also pregnant, was being mistaken for Jackie Kennedy, which made our trips around the downtown area rather unusual. Luckily, we didn't realise that such a misidentification could be dangerous in America. We were still thinking like Australians, to whom the possibility of politically motivated violence was unheard of.

The Kennedy presidency was sometimes referred to as the American version of the legendary Camelot. To me and my family, as Tasmanians, the whole of America seemed somewhat like Camelot. The high standard of living, the availability of good clothing in myriad sizes, the invigorating discussions with our peers, seemingly unlimited support for my research and the availability of a wide range of world-standard newspapers made for a heady mixture.

It was a time of major social change in America. When we arrived in 1959, areas like toilets and buses were still divided into white and non-white sections in the South. The viciously racist Klu Klux Klan was still occasionally active. There was a strong anti-segregation movement, both on the university campuses and throughout society as a whole. We lived about two kilometres from Harvard Square, where the movement against segregation was very much in evidence, and we saw many of the well-known protest singers in action on the streets.

My father's younger brother, from Scotland, was the senior minister at Riverside Church in New York City. It was a block away from the Juilliard School of Music, and this noted institution provided the church choir and occasional soloists. The quality of the music in this wonderful Gothic church was a revelation to us.

Gillian attended a service there at the height of the civil rights riots, protests and street battles. Martin Luther King Junior was the guest preacher and his sermon built on his vision of a new, integrated American people. His famous declamation, 'I have a dream', resounded loudly wherever we were during those times. Looking back, we can see that we lived through one of America's most substantial periods of social change in recent history. It made us think a great deal about the Australian situation.

We became much more international in our outlook. I had Brazilian, Bolivian, Italian and Indian colleagues, among them were Jews, Muslims, Hindus and Christians.

At that time most Australians were poorly disposed to the Japanese nation. Many families had been severely affected by the war in the Pacific and it would take two more decades before the animosity died away.

I, however, was sharing my office with a young Japanese scientist who worked with Bruno Rossi. This seemed quite natural in the university community and there were no difficulties on either side. Another young Japanese scientist, who became a lifelong friend and the 'father' of the Japanese space program, told us about his experiences as a junior officer in the Japanese navy towards the end of the war.

He recounted how, the day before Japan surrendered, he and a colleague 'liberated' (stole) the radar equipment from the last surviving Japanese aircraft carrier. They took it to Nagoya University, where it allowed them quickly to become world leaders in the study or radio waves emitted by the Sun. He said it was touch and go whether they would be shot. Obviously, he was another space physicist who was not risk adverse.

Towards the end of 1961, I was offered assistant professorships at four American universities. After much thought I chose the one offering the lowest salary—the Massachusetts Institute of Technology. Then in 1962 I was headhunted to become a full professor in a new research institute near Dallas, Texas.

Dallas was an up-and-coming city, well known for its oil entrepreneurs and its financial, electronic and aviation industries. It was home to the company Texas Instruments, which had started out in oil exploration then became an extremely successful electronics company.

The three senior shareholders of Texas Instruments wanted to develop an academic tradition in Dallas and they provided a large endowment to establish the Southwest Center for Advanced Studies. Soon after I arrived there, I received huge amounts of government funding for several projects and submitted my proposal for the *Pioneer* satellites (see Chapters 16 and 17).

We bought our first home in suburban Dallas. It was a brand new, four-bedroom house, which had many of the modern conveniences we had only read about. We secured a government backed mortgage at extremely good interest rates. All of this strongly reinforced the Camelot image.

All of our neighbours were young and almost every family had two or three young children. Our two daughters fitted in quickly. Our house was always full of their friends and we unsuccessfully tried to develop in them a liking for Vegemite.

About this time Walt Disney introduced a Tasmanian devil character into his movies and for a while we would occasionally respond to a knock on the door to find a small child asking to see the Tasmanian devil. Our children started school and were soon better able to count in Spanish than in English.

Cecil Green was one of the three founders of the Southwest Center for Advanced Studies. Some years previously he had provided Professor Harry Messel from the University of Sydney with the funds necessary to build the SILLIAC digital computer. This had been modelled on the ILLIAC computer at the University of Illinois and these two were among the first computers ever built. Harry would occasionally drop into Dallas and come storming into my laboratory 'to see what I was up to'.

On one occasion, Gillian and I attended a memorable 'Australian party' at the Green's house in the millionaires' belt of suburban Dallas to celebrate Harry Messel's arrival back in town. We met a wide cross-section of the Texan community, ranging from several Australian war brides to an obnoxious Texan developer. He moaned all night about how Sydney City Council would not permit him to demolish a historic sandstone building in Macquarie Sreet—in order to build an eight-storey hotel. He blamed 'the communistic nature of the government of Australia'. The Australian prime minister at the time was Sir Robert Menzies, whose electioneering was based on strongly anti-communist rhetoric. We found this somewhat puzzling.

Texas, of course, was a bit different from the majority of America. The car park at the Southwest Center for Advanced Studies illustrated this. Quite a number of the technical staff drove what we Australians call utes and the Americans call pickup trucks. Many had gun racks behind the driver's seat and it was common to see three or four high-powered rifles through the back window. On Tuesdays a big glass bottle—of about six litres (1.5 gallons)—and a pile of money was left in the front seats of some vehicles. This was the day for the weekly delivery of moonshine—illegally distilled bourbon that was the 'real man's drink of the southern states of America. In 1963 a large part of Dallas was still under prohibition.

We travelled widely. I was invited to Mexico City to teach the space research group at the University of Mexico how to use the computer program I had developed at MIT.

Gillian, our two daughters and I had a marvellous time there, climbing the pyramids, visiting 16th-century mining towns in the vicinity of Mexico City and the incredible Museo Nacional de Antropología. We would occasionally visit an isolated holiday home in the dunes of a wide, sandy beach on the Gulf of Mexico.

Washington also sent me on a tour of the universities of the Mid-West to spread the good news about the space program. We enjoyed several six-week tours of America and Canada in our trusty Rambler station wagon, which doubled as a sleeping space for all four of us at night. I developed a close association with a research scientist at the Canadian atomic energy authority in Ontario. Our family visited him several times, experiencing both their very cold winters and the sunny, 18-hour days of their summers.

It was a very full life. The telephone connections to Australia improved in 1963, with the installation of the first trans-Pacific telephone cable, and we could speak to our families without them sounding like Donald Duck. Australian friends and family also came to visit. The life of this Australian space scientist—and his family—was very good indeed.

‹ 19 ›

Lunch with the President

Almost immediately an official came to our table and whispered to the mayor and to the philanthropist beside Gillian. The latter gave out a gasped 'no' and slumped sideways against Gillian, who thought he was having a heart attack. He whispered to her, 'There's been an accident. The President has been injured'.

Towards the end of their first term of office, American presidents go into 're-election mode'. They travel the country, meeting, greeting and being very visible. Thus it was announced that President John F Kennedy and the First Lady would visit Dallas in November 1963.

Now, Dallas was a very conservative city that has always strongly supported the Republican Party. Kennedy was a Democrat. That and other factors resulted in there being some hostility to the proposed visit. It was agreed with the city fathers that education and the space program were the only topics that he would discuss in public. A formal lunch was arranged, at which the President would address the city fathers, important business people, educators and senior space scientists. Since I fell into the last category, Gillian and I received invitations.

It was a cold, clear, early-winter's day. The sun was shining brightly, but the chill in the air meant that we were rugged up in thick coats. Entering the venue, we deposited our coats in the cloakrooms.

We found that we were seated with two of the philanthropists who had set up the Southwest Center for Advanced Studies. One was also the Lord Mayor of Dallas. Gillian was sitting immediately next to the other. I started to tell the philanthropists about the progress we were making with our *Pioneer 6* instrument and, in particular, how it would not be flying if it were not for the integrated circuits invented by their company.

Gillian was soon discussing recent acquisitions at the Dallas Art Gallery with their wives. These women were intent on making Dallas a major cultural hub of America and their vision for the city was exciting. There was a real 'feel good' atmosphere throughout the dining room.

I then heard several sirens go past the building. I thought nothing of it; they were a common occurrence in Dallas.

I then heard several sirens go past the building. I thought nothing of it …

After a few minutes the chairman announced that the President had been somewhat delayed and that the caterer would soon serve the first course. He explained that this would allow us to finish lunch before the President needed to leave for his next engagement.

With some surprise we began our meal, but almost immediately an official came to our table and whispered to the mayor and to the philanthropist beside Gillian. The latter gave out a gasped 'no' and slumped sideways against Gillian, who thought he was having a heart attack. He whispered to her, 'There's been an accident. The President has been injured'.

Several minutes passed. Then the chairman of the lunch returned to the podium. He was visibly shaken. He announced, 'The President has been seriously injured and is on his way to hospital. I ask that you all leave immediately'.

We all left, fearing the worst. Gillian went to collect her coat and found the cloakroom attendants glued to a small radio. They were apparently incredibly distressed and almost hysterical. They shouted that the President had been shot!

The babble of discussion by the women attending the lunch about what might have caused the delay, stopped in an instant.

These attendants were from the more disadvantaged section of the populace of Dallas. Almost certainly they would have been Kennedy supporters. They had been listening to the description of the motorcade on the radio. From them Gillian learned that there had been several rifle shots and that the President appeared to have been seriously injured by more than one gunshot and was feared dead. We drove home. En route the radio announced that the President had died.

Over the following days the whole strange tale unfolded. First, there was the surprisingly easy capture of the presumed assassin, Lee Harvey Oswald. Then, at the very time that Oswald was being transferred to another prison, Jack Ruby, a person of doubtful reputation, was allowed in the police station with his revolver, where he somehow managed to be close enough to Oswald to shoot him at point blank range. Of course, the television cameras were pointing in the right direction to broadcast it all nationwide.

Several weeks later Gillian was shopping at the small supermarket near our home. There, sharing the aisle with her was Marina Oswald, the frequently photographed wife of Kennedy's presumed assassin. They nodded to each other.

We had been in America for just four years and experienced many things that had been unimaginable when we left Tasmania, but after being close to the assassination of a president, our personal Camelot was shattered. I began to look seriously for a suitable position back in Australia.

‹ 20 ›

There but for the grace of God …

The purpose of our instrument was to count the cosmic rays coming from different directions in space. The electronics were supposed to count just as we do—1, 2, 3, 4 and so on. We now found that it was counting 1, 2, 5, 6, 8, 9 and so on.

With great fanfare, America launched the *Hubble Space Telescope* in 1990. It carried the most powerful telescope ever built by mankind and would see further into the universe than would be ever possible from the surface of Earth. For obscure reasons astronomy has always been a good vote catcher and NASA was looking forward to *Hubble* reviving its failing appeal to the American taxpayer.

After the fanfare, there was silence. Then rumours began to circulate that there was something seriously wrong with *Hubble*. Then the awful truth emerged: the mirrors that do all the clever work in the telescope were the wrong shape. America had a $1.5 billion space lemon.

The mirrors that do all the clever work in the telescope were the wrong shape. America had a $1.5 billion space lemon.

All around the world the people who had built research satellites and spacecraft paused and reflected on their own lemons, or near lemons, and were relieved that there had been no fuss about them. I, along with everyone, thought, 'There but for the grace of God ...' I could easily have been in the position of the people who built *Hubble*.

The history of space research is full of lemons. Space research is a high-risk business. The following are just two occasions that illustrate what it means to 'do a *Hubble*'.

The first is an example from the really early years of the Space Age. Then, as now, you would begin building a satellite instrument by writing the specification. This is just like the specification in the back of the instruction book for a hi-fi or a PC, except that it is very long. It often comes in several volumes. It tells you how the thing will be built, how it will be tested and what the results of those tests will be. If the thing is being built by a company, the specification has an extremely important role. Only when the instrument 'meets the specification' is the company paid.

The instrument in question in this case was designed to measure the solar wind. As discussed in Chapter 14, this wind consists of protons and electrons that stream away from the Sun at about 450 kilometres (280 miles) per second. The solar wind has a strong impact upon the Earth's magnetic field over the polar regions, which does very nasty things to the radar signals that the military were using to detect intercontinental ballistic missiles which might be heading their way. Understandably, they were very keen to know more about the wind.

This was far from easy to do, because the strong sunlight coming from the same direction as the solar wind would actually appear as a solar wind a million times greater than the real thing. So a very clever instrument was designed. Having built a 'laboratory version', the scientists wrote the specification for the construction and testing of the flight instrument by a top electronic company. One very important test required that it should be pointed at strong ultraviolet light to make certain that its clever parts would cause the instrument to ignore the light. The test was run, and the instrument passed with flying colours.

The great day finally came. The satellite roared into orbit and the instrument was turned on. Then ... shock, horror! The solar wind instrument was being 'blinded' by the light from the Sun. The clever bits were not working at all. The instrument was utterly useless.

How could this be? The instrument had been tested and shown to be totally insensitive to light.

The scientists who had designed the instrument asked for all the test results again. They analysed the readings that the technicians had made and agreed that the instrument had not responded to the bright light used in the tests. They then asked to see the test equipment itself. They were shown all the electronics, meters and oscilloscopes.

Then one of the scientists asked what he thought was a stupid question: 'Where is the vacuum chamber used to replicate the vacuum of space?'

'What vacuum chamber?' the engineers said. 'The specification didn't say anything about a vacuum chamber or a vacuum.'

'But surely that was obvious', the scientists said. 'Everyone knows that you need to test satellite instruments in a vacuum'.

The engineers agreed, replying, 'That's right, so when your specification didn't ask for vacuum, we assumed you knew what you were doing and that it was important to do the tests at normal air pressure'

That was the end of the matter. A simple sentence had been left out of the specification. The absence of that sentence led to the acceptance of a faulty instrument for space flight. As a consequence, the satellite was a complete failure.

> *A simple sentence had been left out of the specification ... As a consequence, the satellite was a complete failure.*

We all knew about this lemon. We took great care to check and double check our specifications, but Murphy's law still managed to strike my instrument another way.

In the early days of space research we had to build all our instruments very quickly. The four different companies building the various bits of our *Pioneer 6* instrument would always deliver a few days late and, although we had scheduled a number of comprehensive tests in our laboratory, we never had the time to properly evaluate them.

To our great relief, it was decided that all the experimenters on *Pioneer 6* would be given back their instruments a month or so prior to 'blast off' for a final calibration.

It was emphasised that we were not to change the electronic wiring in the slightest and, to be safe, NASA sent someone to watch over it. We took this chance to put our instrument into round-the-clock tests.

After several days a technician ran into my room to say something very bad happened to our instrument when it was cooled to about 5 degrees Celsius (41°F). It had forgotten how to count properly.

The purpose of our instrument was to count the cosmic rays coming from different directions in space. The electronics were supposed to count just as we do—1, 2, 3, 4 and so on. We now found that it was counting 1, 2, 5, 6, 8, 9 and so on.

Our first reaction, and hope, was that there had been some permanent change to the electronics so that it would always count in this new manner. But no, when we warmed the instrument up, it went back to counting normally.

We now had a major problem. When the instrument said it had seen nine cosmic rays, did it mean six or nine? If it said 27, did it mean 27 or 53? Our instrument had been designed to detect a 0.1 per cent difference in the cosmic radiation flux—with this type of error, we would be lucky to achieve an accuracy of 50 per cent. Unless we could overcome this problem, we would never be able to trust our results. We would have a very expensive lemon and, worse still, might never be given another ride in space.

Our first hope was that if we could measure the temperature accurately, we would be able to decipher the results. A few investigations destroyed that hope. We were in real trouble.

My team and I considered the situation. We kept up our round-the-clock series of tests. The guardian of the instrument quickly felt the strain and decided to restrict his vigilance to eight hours a day. We took the obvious precaution of not telling him that we had a problem. To fill in his time, he tried to seduce one of the secretaries in my group.

After consideration of all the data and examination of the electronic circuits, we decided that there was no way we could remove the fault using the data alone. As happens so often in science, we could see a way around the problem—but there was a catch.

The following midnight, the conspirators met at the laboratory—the scientists, the head engineer and our most experienced technician. We dismantled the electronics, then carefully—against every instruction given to us by NASA—we cut one wire, and replaced it with a 5-millimetre wire running to another point in the electronics. We had worked out that this would give us the information we needed to be able to tell exactly which counting scheme our electronics were using.

We quickly reassembled the instrument and put it back into the cold chamber. We resumed our tests, but now with the secret goal of verifying that we could always decipher the scrambled counting from the instrument, no matter what. A day later we knew we had succeeded.

Several days later we handed the instrument back to its guardian, who returned it to California to bolt into the satellite for the last time. We assured him honestly that it was in excellent condition and extremely well calibrated for flight. We never did explain the lengths to which we went to achieve that calibration.

We decided that this information should be issued on a 'need to know' basis and that NASA had absolutely no need to know about what we had been up to. We didn't want to give them unnecessary headaches.

Three months later our satellite was 25 metres (80 feet) above ground, screwed on to the pointy end of a 'thrust augmented' *Delta* rocket. There was a platform around it from which the NASA technicians could lean into the satellite to make the final tests of the instruments. The tests were being run 24 hours a day.

One morning we went into the NASA project manager's office to be told, 'We are taking your instrument off. It has had a massive failure. We cannot take the risk that it could cause damage to other parts of the satellite'.

'Like hell you are!' was the polite version of my reply. We asked for the evidence of malfunction. We were told that, during the technicians' last hands-on tests of the instruments, they had inserted a polarised plug into a connector in our instrument. (Here 'polarised' means that the plug had been very carefully designed to make certain that it could not be put in upside down.) This test showed that all the test voltages were completely wrong. Everything was completely scrambled. It certainly looked like something catastrophic had happened.

'What does the data from our instrument look like through the telemetry?' I asked. In other words, what data were being transmitted by radio from the satellite.

'Oh, that looks perfectly normal', I was told. On the computer printout from the telemetry, sure enough, our instrument was merrily counting cosmic rays. Clearly, no one had told it yet that it wasn't supposed to be working.

I raised hell.

The NASA project manager was far from pleased, but he had to agree that there was some doubt that the instrument was faulty. He said he would have it tested again, but if the tests were faulty, it would be taken off without further discussion. I insisted that my team were there while the tests were being made. This was 'against the rules', but we were allowed. We raced to the launch pad and raced up the stairs to the pointy end.

The technician had done the supposedly impossible. He had put the foolproof plug in upside down.

The tests were already in progress. The technician said the voltages were still completely scrambled. 'Let me see what you have done', my engineer shouted. He leaned down into the satellite, and examined the test plug. 'You've got it in upside down', he said. He reversed the plug and suddenly the instrument tested as being in perfect health. The crisis averted, the countdown continued.

The technician had done the supposedly impossible. He had put the foolproof plug in upside down. Without our insistence, that would have been the end of our flight and the whole scientific mission would have been seriously affected. Murphy's Law almost beat us that time.

‹ 21 ›
Blast off

We all knew, deep down, that sooner or later there was going to be a big public exposé of a space experiment that went wrong. We hoped that it would not be us and didn't let this sword of Damocles stop us.

By 10 December 1965 *Pioneer 6*, aka *Spudnik 1*, was ready to go into space. However, Cape Canaveral was not ready for it. Cape Canaveral operated like a railway station on a single line. The rockets were scheduled to go off one after another and we were scheduled to go after the penultimate flight of the *Gemini* spacecraft before mankind went aloft in it. However, the Moon, the Sun and union power would have their say, too.

Launching a satellite to Venus was greatly influenced by the phase of the Moon as well as the time of year and of day. We had a short launch window of only about twelve minutes each day at 2.40 am for just five days. To further complicate matters, two of those days were during Cape Canaveral's Christmas shutdown.

The unions made it very clear—space flight was extremely important, but the Christmas break was even more so. *Pioneer 6* would only have three days in which to launch. Miss that window, and we would have to wait months.

The unions made it very clear—space flight was extremely important, but the Christmas break was even more so.

These facts we knew well enough and we knew *Gemini 6* would be flown a week before our launch window opened, so we didn't think there would be a serious problem.

The countdown for *Gemini 6* started. It reached minus four seconds and the rockets sprung into life. Fire spewed out exactly as it should and the countdown reached 'lift off'.

At that point in, the rocket should have slowly and majestically lifted off. Perversely, the engines stopped. There was just a cloud of steam and a bloody big rocket sitting where it should not have been. All on nationwide television. Everywhere, people like me were thinking 'Why?' and 'Will it blow up?'

NASA was, of course, shitting itself.

They had been having a number of bad years. The Soviet Union beat them at putting a satellite into orbit. Soviet cosmonaut Yuri Gagarin was the first human into space, years ahead of John Glenn. The Soviets were the first to land a satellite on the Moon. The American satellites weighed tens of kilograms, the Soviets' hundreds. NASA had become the butt of jokes, such as:

Question: what does NASA stand for?
Answer: Not A Serious Attempt.

Now there was a reluctant rocket sitting on the launch pad, chock full of kerosene and liquid oxygen. If it blew up, it might take half of Cape Canaveral with it.

Ten horrible seconds went by, then the telemetry told NASA that everything inside the rocket had turned off. Only then did the insiders know that the whole thing was not a big bomb.

As is normal in any government enterprise, a committee was established to find out why the rocket had decided not to go.

It was announced that a piece of plastic wrapper had been in one of the fuel tanks and had gummed up the works of a fuel pump. The grapevine within the space community reported that it was someone's lunch wrapper.

A piece of plastic wrapper had been in one of the fuel tanks ... and had gummed up the works of a fuel pump. The grapevine within the space community reported that it was someone's lunch wrapper.

Having removed the wrapper and refilled the rocket with fuel, *Gemini* was ready to go. So on a beautifully clear winter morning, we stood on top of a large hanger, a couple of miles from the big Atlas rocket that was to take *Gemini* into orbit.

The countdown blared out from loudspeakers all around us. This time the rocket left on cue. It seemed to climb so slowly at first, although it was steadily accelerating.

It was a magnificent sight.

Two days later, at 2.30 am, we stood shivering on a road that went through the marshes of Cape Canaveral. This time it was our countdown. The countdown had already stopped once due to some problem. Now, at T minus ten minutes, it stopped again. This time it stayed stopped. Soon there was the inevitable announcement that the launch was cancelled for that night.

The next night we were out there, shivering again. This time the countdown was flawless and we reached T minus four seconds. The engines ignited. I had never seen a night launch before and was not ready for the intense brightness of the flames. I stepped backwards and fell three metres into the water-filled ditch beside the road. It was a rather poor way to celebrate the launch of my first satellite.

We drove to the blockhouse, where an entire large wall was taken up by an electronic map of the world, showing the rocket's position like a scene from a James Bond movie. There were lots of other screens that gave graphs of speed, acceleration and how all the navigation instruments were working. There was a good feeling in the room; clearly *Pioneer 6* was going 'right down the slot'.

After a minute or so, someone ran up and said, 'You guys are all Americans, huh?'

To which I answered, 'One American, one Australian and one Indian'.

The guy went berserk. 'Out, out, out. Don't you know this is a highly classified facility?'

I tried to explain that we had an experiment on the satellite, and that it seemed reasonable enough that we should be able to see it being tracked. This produced a barrage of swear words and lots more 'outs'. We left, secure in the knowledge that all three stages of the rocket had fired and that it was a perfect orbit. At long last we had an experiment in space.

Satellites are switched on rather slowly. The engineers send radio signals that start one bit after the other. It was about two days before our instrument was turned on—and it worked perfectly. Several days later there was a little explosion on the Sun and a lot of cosmic rays where sent out from it. Our detector told us in great detail how quickly they arrived and the direction from which they came. Our little instrument was doing all we had expected, and much more.

Yet there was this uneasy feeling. Several times we had been in serious technical trouble. In one case, exactly as happened with the *Hubble* telescope 25 years later, we had accepted work performed by our subcontractor that we later found to be faulty.

Of course, in the gung ho 1960s, we all accepted that these things could happen. As we said, 'When no one has done it before, you will not get it right every time', or 'The only person who has not made a mistake in space experimentation is the person who has not done any'.

Yet we all knew, deep down, that sooner or later there was going to be a big public exposé of a space experiment that went wrong. We hoped that it would not be us, and didn't let this sword of Damocles stop us. It was just another risk to accept, along with the cussedness of nature, pedantic bureaucrats and the vagaries of the so-called space race. It went with the territory.

‹ 22 ›

Pioneer 6: an intimate view

Once away from humans, in an almost constant temperature and with no vibrations, bumps or sundry disturbances, the instruments were extremely reliable. We kept calibrating them and examining the results for hiccups, but the answer was always the same: ' I'm okay, now that I'm away from you lot'.

The NASA public relations people called *Pioneer 6* a 'bargain basement' satellite. They said that it had cost a mere $15 million in 1965 dollars—much cheaper than many of the other satellites launched in the 1960s.

Nevertheless, *Pioneer 6* was true to its name. It carried six different experiments, each of which revolutionised our understanding of the Sun and the space between the Sun and the Earth. The spacecraft itself pioneered the way that interplanetary spacecraft would be built in the future, so let me give you a short tour of the bird.

Pioneer 6 looked like a large round rubbish bin, with a thick stick poking along its axis (see image on front cover). There were three long arms hanging out from the circumference of the rubbish bin and the cylindrical portion contained nearly all of the experiments as well as the spacecraft hardware, such as radio transmitters and receivers.

> Pioneer 6 *looked like a large, round rubbish bin, with a thick stick poking up its axis.*

Now comes the clever part.

In the early 1960s, we were already using solar cells to power the spacecraft and the experiments. However, we hadn't learned all the tricks necessary to keep the solar cells pointing at the Sun all the time, nor were we able to keep radio antennae pointing back at Earth from halfway across the solar system. *Pioneer* was designed to avoid the necessity of doing either.

The cylindrical spacecraft was set to spin at one revolution per second, with its axis at right angles to the plane of the Earth's orbit about the Sun. Careful adjustments of the distribution of mass in the spacecraft meant that it would remain spinning like that forever. As the whole of the cylindrical spacecraft was covered with solar cells, it received a constant flow of electrical power. No pointing and no further adjustments were necessary.

Then there was the problem of sending the data from the six experiments back to Earth from clear across the solar system.

As the months went by, *Pioneer 6* would move further and further away from Earth. After six months it would be 50 million kilometres (30 million miles) away; after two years 180 million kilometres (110 million miles). As a result, its radio signals would grow progressively weaker and harder to understand.

Today spacecraft use parabolic antennae—similar to the satellite dishes we use for pay television—to send their radio energy back towards Earth. Without the necessary technological know-how to keep an antenna pointing back at Earth, *Pioneer 6* used a simpler approach. The 'stick' below the rubbish bin was a special type of radio antenna that sent all the radio power into a thin disk at right angles to the stick. Because the antenna was always at right angles to the Earth's orbit, the Earth was always in line with this thin disk of radio energy, no matter how far away the spacecraft was. This was an example of the 'set and forget' technique—once the spacecraft was spinning as intended, the radio energy would always be sent back to Earth without further adjustment.

To capture the signals from the *Pioneers* and other interplanetary wanderers, NASA established three Deep Space Facilities—in California, Canberra and Spain. Because the signals would always be very weak, very large antennae were needed—the larger the antenna, the more signal it will capture.

When *Pioneer 6* was within 50 million kilometres of Earth, 25-metre (80 ft) diameter antennae were used. Beyond that distance, we changed to really big antennae, 72 metres (235 ft) in diameter.

When *Pioneer 6* was close to Earth, the combined effects of the directional antenna on the spacecraft and the big parabolic antennae on Earth allowed the spacecraft to send data back to Earth at 64 bytes per second—that was a big deal back then.

As the spacecraft receded from Earth, the signal grew weaker and began to be confused with the electronic noise that is always present in space and in radio receivers. The problem is similar to having a conversation in a crowded cocktail party. There, the trick is to speak more slowly.

That's what *Pioneer 6* did too. Progressively it decreased its rate of data transmission, first to 32 bytes per second, then 16, and so on until it was transmitting at only 1 byte a second. We didn't complain—we were seeing what was happening on the other side of the solar system.

A slow data rate was a minor inconvenience compared to the value of the data we received.

Now to the inside of the rubbish bin. Around the circumference were the six experiments, peering out into space through a 'belly band'. Nearer to the centre were the power supply equipment and batteries, the radio transmitters and receivers, gas bottles for the 'attitude control' system, and all the other stuff needed to operate the spacecraft.

Following a nationwide competition, six different laboratories were chosen to design and supply the six experiments that made up the scientific payload.

The University of Chicago provided an instrument to measure the chemical composition and intensity of the cosmic radiation.

My friends at the Massachusetts Institute of Technology provided an instrument to measure the solar wind, similar to the one I had seen on my first morning at MIT.

The Ames Research Center of NASA supplied a different type of instrument to measure the solar wind. In those early days it was necessary to 'back up' and use several different types of instrument to measure the most important properties of space.

There was an experiment from Stanford University to measure the effects of space upon the propagation of radio waves.

NASA's Goddard Space Flight Center supplied a key instrument to measure the three components of the magnetic field in space. Despite our best efforts, the spacecraft had a weak magnetic field of its own, so the magnetic sensors were on one of the long arms to position them as far away from the spacecraft field as possible.

Then there was my experiment, supplied by my research group at the Southwest Center for Advanced Studies. (Plate 4a)

All the cosmic ray experiments flown in space until that time had not worried about where the radiation in space came from. We were pickier. We would measure, with an accuracy of 0.1 per cent, the manner in which the radiation intensity depended upon the direction from which it came.

Was there more from the Sun or from the Milky Way? As the spacecraft spun about its axis, we computed the direction the instrument was pointing each time it saw a cosmic ray.

The six instruments continually measured the properties of space and stored their results ready for transmission back to Earth. Every 30 seconds or so, the spacecraft sent an electrical signal to our experiment asking it to supply its latest readings. The experiment responded by sending the five binary-coded 'words' that contained all of our data. The spacecraft's electronics added error-checking digits and the radio transmitter then sent them all to Earth.

The next experiment would then provide its data and so on, until it was our turn again. The spacecraft also recorded 'housekeeping' data, which measured about a hundred different things, such as temperatures, voltages and currents, which allowed the engineers to monitor the health of the whole spacecraft.

Forty years ago, electronics were not as robust as they are today. Despite the greatest care in manufacture, the very best scientific instruments would be relatively unreliable in the laboratory. All experimenters became very skilled in the diagnosis of electronic problems and at fixing them. Seldom would a week go by without needing to tweak some control knob, replace something, blow the dust off or just give everything a good shake to overcome problems in plugs or other parts. In the army we had had the same sort of problems with our radio transceivers, but were a little more direct in how we dealt with them; often our first solution was to 'give it a good kick in the guts'.

On several occasions, new 'discoveries' were trumpeted as having a revolutionary impact on physics, only to find later that they were due to a faulty instrument.

We were building instruments that we wanted to work for years without our tender attention. This caused us great concern. In particular, we wanted to know if changes in our instruments would fool us into thinking that something weird and wonderful had happened to the properties of space. An experimenter's greatest nightmare is that an unstable instrument will lead him to 'discover' something that isn't there—this has happened many, many times.

On several occasions, new 'discoveries' were trumpeted as having a revolutionary impact on physics, only to find later that they were due to a faulty instrument. So we were very much aware of the risks to our sanity and reputations if our instruments started playing silly games of their own once they were out in space.

To help to solve the problem, we all built calibration systems into our experiments. In our case it was a very weak radioactive source, which would produce a constant number of alpha particles, of precisely known energy. A radio signal would then be sent from Earth that caused our experiment to measure the alpha particles coming from the radioactive material. If these electrical signals were the right size, and there was the right number each second, then we knew that our gismo was doing its stuff.

Despite our fears, we soon learned another fundamental truth of space exploration: provided an instrument doesn't fail in the first several weeks—suffering 'infant mortality', as we called it—it will keep working well for a very long time.

The most hazardous time for a space instrument is actually the period before it is launched into orbit. Bad things do happen to experiments then, all caused by the proximity to humans. This 'finger trouble' might include people wiggling cables, electrostatic discharges from our bodies and basic actions like turning the spacecraft on and off.

Once away from humans, in an almost constant temperature and with no vibrations, bumps or sundry disturbances, the instruments were extremely reliable.

We kept calibrating them, and examining the results for hiccups, but the answer was always the same: 'I'm okay now that I'm away from you lot'. Our instruments didn't fail once in the five or so years before the spacecraft was shut down.

The antennae of the Deep Space Network recorded the weak radio transmissions coming from *Pioneer 6*. The next problem was disentangling all the data.

The data comprised a single stream of 'bits', each bit representing either a one or a nought. (Modern computers use 'bytes' which are comprised of eight bits.) There would be 30 bits for me, 48 for the University of Chicago, the same for MIT and so on, every 30 seconds, but which were my 30 bits?

First the data stream was recorded on a computer tape and sent to the Ames Research Center in California. There a computer would look for a 'synchronisation word', which would tell it where to start counting the separate sections of data. It would then write six different computer tapes, one for each of the experimenters.

With scientific and organisational reputations at stake, each group received only their own data and absolutely nothing from the others. The experimenters would exchange data later, but not before it had been checked extremely carefully. Remember, we were all scared stiff that our experiment might play up and wanted to be the first to know.

This was particularly true in our case, since our *Pioneer 6* instrument had forgotten how to count properly. We knew how to unscramble our data, following our clandestine last-minute changes, but we did not intend to advertise this. We gave our data to the other experimenters only after it had been unscrambled and thoroughly checked for other problems.

Over subsequent years I learned that many other early experimenters experienced similar problems to ours. Some of them were still trying, unsuccessfully, to unscramble their data twenty years later.

A common issue was what engineers call 'radio frequency interference'. On a spacecraft this arises when you have very sensitive electronics sitting a foot away from a powerful radio transmitter, which can be heard 180 million kilometres away on Earth.

It's a bit like trying to have a whispered conversation when sitting in front of a rock band. The instrument would pick up the radio signals and you'd have enormous trouble. Performing space experiments in those hands-on days was never easy.

Our second instrument, identical to the one on *Pioneer 6*, was flown a year later on *Pioneer 7*. While the earlier mission had gone to the vicinity of Venus, *Pioneer 7* went outwards from the Sun to the vicinity of Mars.

Even before this was launched, NASA had granted us 'rides' on five more spacecraft: two on the Interplanetary Monitoring Platforms (IMP), which were launched into highly elliptical orbits about Earth; then three more *Pioneers*, for which we designed a more complicated instrument based on what we had learned from *Pioneer 6*. (Plate 4b)

Even though we were given these rides when rockets and satellites were far from reliable, the scientific gods mostly smiled benignly on us; the rockets for *Pioneers 6* and *7*, *IMPs 4* and *5*, then *Pioneers 8* and *9* all worked perfectly.

Then the gods served us a small reminder of their power. My last spacecraft ended up in the Atlantic.

By the time of our last mission, we were in a situation that we had never thought possible back in 1963. For almost a year we had five spacecraft spread out around the solar system.

At one point *Pioneer 6* was 180 million kilometres (110 million miles) away behind the Sun, *Pioneer 7* ninety degrees away, and *Pioneers 8* and *9* each about 30 million kilometres (18.7 million miles) from Earth, while *IMP 4* was tethered to Earth. (Figure 4)

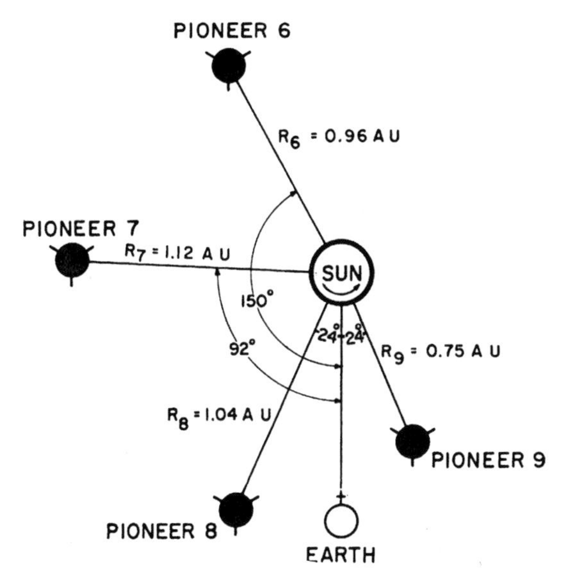

Figure 4: The location of the author's experiments in the solar system in 1969. The data from all these spacecraft were received by the deep space reception facilities.

We had a grandstand view of what the Sun was doing. When there was an explosion on the Sun, we could see that the radiation reaching the five satellites was very different. We were able to study the properties of interplanetary space in a manner that was inconceivable ten years previously.

That was the story of those early days of space research. If we had started by asking for what we ultimately received, we would have been regarded as totally crazy. However, by taking a few risks and progressing one step (or spacecraft) at a time, we achieved more than we had ever dreamed possible.

‹ 23 ›

The excitement of discovery

Three days later the first magnetic tape arrived and was hastily processed on our mainframe computer. We had eight hours of data, more than enough to show that our instrument was working exactly as intended. It was Christmas Eve, and this was a magnificent Christmas present.

The morning after the launch of *Pioneer 6*, we visited the NASA project manager's office at Cape Canaveral. The spacecraft engineers were happy.

The satellite had entered its parking orbit with commendable accuracy. The satellite operations crew had performed an important manoeuvre, which oriented the spacecraft to gain maximum power from its solar cells. The radio transmitters and receivers were working well and all of the 'housekeeping' data was well within specification. The rocket used to position the spacecraft in its interplanetary orbit had then been ignited and worked to perfection.

Pioneer 6 was looking very good, indeed.

The experiments were then left to 'soak' in the vacuum of space for several days before they were turned on. Many of the experiments used quite high voltages—ours used 1300 volts in some of the detectors—and there had been cases where a high voltage had been turned on before all the atmospheric gas in the instrument had escaped into the vacuum of space. An electric arc had formed in the residual gas and ... *ZAP*, the experiment had been killed, stone dead. Obviously, we were content to wait for a few days to avoid that sort of fate.

Returning to Dallas, we waited patiently until we heard the good news by telephone: our experiment had been turned on and by all indications it was working properly. Four days later the first magnetic tape arrived and was hastily processed on our mainframe computer. We had eight hours of data, more than enough to show that our instrument was working exactly as intended. It was Christmas Eve and this was a magnificent Christmas present.

Rapidly the news became even better. We knew that large explosions on the Sun released a burst of radiation that would arrive at Earth some ten to twenty minutes later. We also knew that there were many much smaller explosions on the Sun, but there were good reasons to think that the strong solar magnetic fields might prevent a small burst of cosmic radiation from reaching Earth. This sort of behaviour is called a threshold effect and is a common feature of natural systems.

Lightning is a good example of a threshold effect. You only see the bright lightning bolts when the voltage in the atmosphere reaches a critical value. Nothing is visible until then, although a powerful electrical voltage is often present long before it gets big enough for a lightning bolt to occur.

The conventional wisdom in those days was that it was only possible for solar cosmic rays to reach Earth if there were enough of them to jointly punch a hole in the strong magnetic fields surrounding the sunspot (like a rugby scrum pushing the other team out of the way). If so we would never see the small bursts of radiation, because the cosmic rays would not be able to force their way out. The absence of a threshold effect for the radiation bursts from the Sun would be an important discovery and would tell us a lot about our star and exactly how it accelerates cosmic radiation to the high energies it achieves.

The Sun didn't leave us in suspense for very long. On 30 December our instrument detected a tiny burst of cosmic radiation, shortly after radio and optical telescopes observed a very faint 'solar flare'. The cosmic ray burst was a thousand times smaller than any of the bursts previously seen by any instruments, either in space or on the Earth's surface. Our experiment had emphatically delivered its first result—there is no threshold effect. (Figure 5)

The little burst of radiation gave us another important result as well as verification that the instrument was working properly. The instrument was designed to measure the cosmic radiation coming from four different directions in space. When the little burst of radiation first reached *Pioneer 6*, it was coming from the direction of the Sun. Over the next half hour the direction slowly changed by about 90 degrees. Then, all of a sudden, the instrument told us that the same amount of radiation was reaching it from all directions in space. This told us that there were quite strong magnetic fields in interplanetary space and that they were varying in direction from minute to minute and hour to hour. This was a very important result, and we lunched out on it for quite a while.

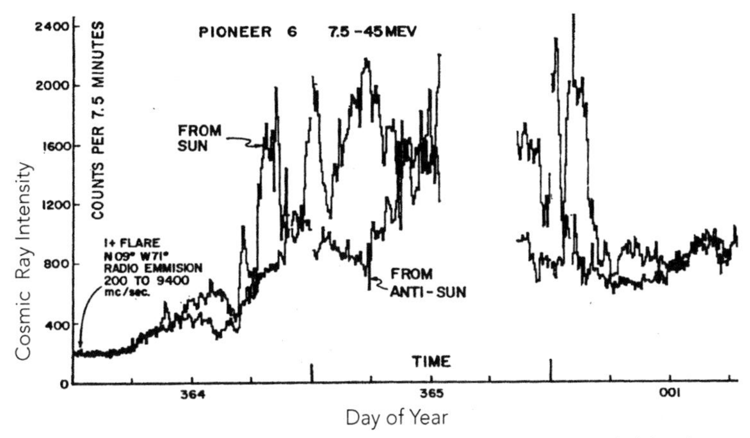

Figure 5. The first burst of cosmic radiation from the Sun recorded by the author's instrument on Pioneer 6. The diagram shows that the radiation coming from the direction of the Sun and from the opposite direction (anti-Sun), were very different.

Nevertheless, we were still concerned that the instrument might tease us by working well for a few days, before stopping suddenly or starting to send undecipherable gibberish. Despite all the tests in the laboratory, there was absolutely no way to anticipate all the stuff-up factors that might occur in space: the rotation rate of the satellite was changing slowly and that might confuse our electronics; our power supplies might play up; or the integrated circuits might play tricks, just as they had done before launch. The other experiments might radiate signals that would confuse our electronics. It was a very nervous time.

The key feature of our instrument was that it needed to know where it was pointing at all times as the spacecraft rotated through 360 degrees. This would have been simple enough if the spacecraft rotated at a known and unchanging rate, but it did neither. The rotation rate after launch could be up to 25 per cent different from the design value and it would change over time, sometimes quickly, sometimes slowly.

We designed a 'digital divider' to allow for these vagaries—at a time when digital electronics were still in their infancy. If this didn't work perfectly, we were in big trouble. Working properly 99.99 per cent of the time was not good enough; we needed 100 per cent, for six months at least, or the experiment would be a failure. This was a big ask in space science in those days.

The first little burst of cosmic radiation dispelled most of our doubts. The digital divider did a marvellous job. The internal checks were perfect. The direction sensing system worked perfectly for hours, then days on end. By New Year's Day 1966, we knew that the greatest risk to our data, and our reputations, had been removed.

There were, of course, many more things that might yet go wrong, but this was the one we feared the most. With that hurdle behind us, we could start enjoying the data, the scientific discoveries and the wonder of seeing things that had never been seen by mankind before.

We carefully studied our results and were surprised to see that the radiation from the Sun kept coming from strange directions that we had never anticipated. We checked these directions against the magnetic measurements being made by the magnetometer on *Pioneer*. There was good agreement.

We went to lunch at the canteen pondering the strange results. I chose spaghetti bolognaise from the menu and, as I sat looking at the spaghetti on my plate, it suddenly occurred to me that our experiment was telling us that the magnetic fields in space must look like that. The strands of spaghetti were like magnetic 'tubes' of force and, just as the spaghetti had become intertwined by the cook stirring the pot, the magnetic tubes had been mixed by the effects of turbulence in the solar wind. So the 'wet spaghetti model' of the interplanetary magnetic field was born. (Figure 6)

Scientists frequently use 'models' based on everyday life to help them to understand the complexities they see in nature. It helps us to get our mind around strange results or difficult concepts. We published scientific papers describing the 'wet spaghetti model' and used it to discuss our results. Other scientists used it. In fact, our wet spaghetti model provided a good description of the interplanetary field for at least twenty years until it was superseded by a mathematical description invented by the theoreticians. Personally, I preferred the simplicity of our spaghetti model.

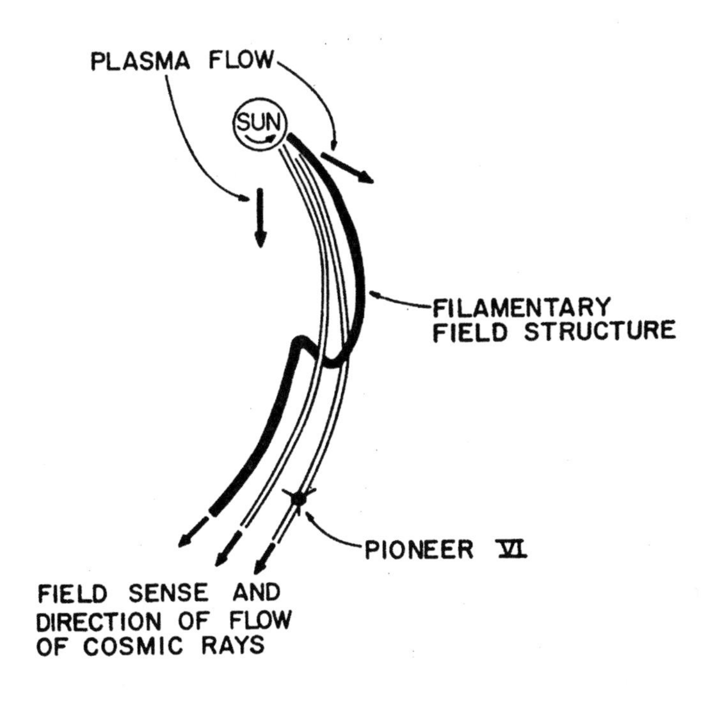

Figure 6: The inferred configuration of the interplanetary magnetic fields in space, based on the cosmic ray measurements made by Pioneer 6. *(popularly known as the wet spaghetti model of the interplanetary magnetic field).*

The months went by. The Sun kept spitting out cosmic rays. We kept analysing the *Pioneer* results. Frequently a colleague would run into my office in a state of excitement and say, 'Hey, look at this'. It was a rewarding time; we were seeing new results. We were sharing exciting ideas and data with the other experimenters on *Pioneer 6*. It was great to be a scientist.

A special *Pioneer 6* scientific conference was held in Washington six months after its launch to report the discoveries made. It was held in a large conference room, but was full to standing room only. It was a conference no space scientist wanted to miss.

The big room gave me a serious problem, though. It meant that a very large slide projector was needed, with a very bright and hot projection lamp. As I stood up and described our results, I would ask for the next slide. It would appear then, after about fifteen seconds, the image would turn brown and become highly distorted as great gobs of photographic emulsion melted and ran down the glass plate. This caused great hilarity and some suggestions that I had done it purposely to prevent people looking at our results carefully.

So our entry into space research was going very well and it soon became even better. Up until then, all our knowledge about the solar system had been obtained from instruments on Earth, or on satellites orbiting the Earth. We only knew what the Sun and solar system were doing when observed from a single point in space.

There was a big unanswered question: what would we see if we could make simultaneous observations from a number of points in the solar system? For example, when we saw a burst of cosmic radiation from the Sun, what would we see if we were also observing from near Venus or behind the Sun?

We were soon to find out.

In August 1966 we launched *Pioneer* 7 from Cape Canaveral. By then *Pioneer 6* was 130 million kilometres (80 million miles) away from Earth and from *Pioneer* 7. Being aware of its serious responsibilities to space scientists, the Sun produced a spectacular burst of radiation a month later.

That is, it appeared spectacular from *Pioneer* 7 but—initially at least—very small at *Pioneer 6*. Over the following days, the intensity at *Pioneer* 7 dropped steadily, while at *Pioneer 6* it rose steadily at roughly the same rate. We were beginning to understand how lumpy the radiation could be in space—very high in some parts, and low elsewhere.

Our understanding of this lumpiness continued to grow as *Pioneers 6* and 7 receded from Earth and with the launch of *Pioneers 8* and *9*, so that by 1969 we had the flotilla of spacecraft (illustrated in Figure 4) providing us with simultaneous observations throughout the solar system.

Now we could really appreciate the lumpiness and what it would mean for astronauts if they were ever to go to Mars,say. Figure 7 illustrates what we could see. There were by then so many spacecraft in 'deep space' that the receiving stations could only acquire data from our more distant spacecraft every second day, but this was more than enough to see the great difference throughout the solar system.

Note that this graph uses a 'logarithmic scale'—the intensity from bottom to top increases by a factor of one million. When the flare occurred, the intensity at *Pioneer 9* was a factor of hundred times greater that at the other spacecraft. Over the subsequent week, the intensity at *Pioneer 9* became the lowest, while that

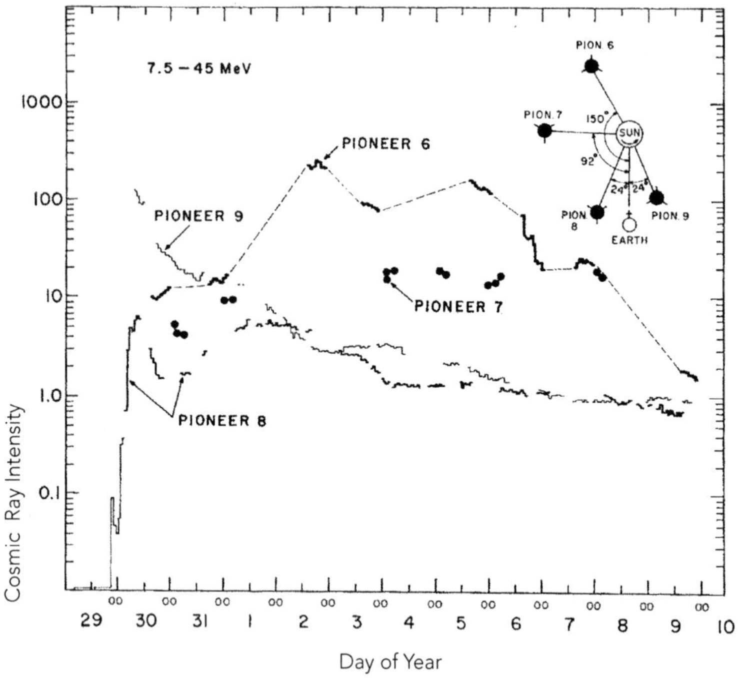

Figure 7: A very large burst of cosmic rays seen by four widely separated spacecraft. The radiation scale is a "logarithmic" scale; representing a million-fold increase from bottom to top. The graph shows that the radiation was about 300 times more intense at Pioneer 6 than at Pioneer 8 (near Earth) and that the radiation intensity remained very high for more than 12 days.

at *Pioneer 6* on the other side of the solar system was a factor or three hundred times greater.

Previously we had known that the radiation would be lumpy, but we didn't have the information that would allow us to estimate by how much. The *Pioneer* flotilla gave us that information. Now we could see that an astronaut en route to Mars might receive a hundred or more times intense radiation than near Earth, and that, as in the case of *Pioneer 6* in Figure 6, the high radiation intensities might last for a week or more.

This was not cheerful news for the manned spaceflight people, but good engineers love factual data and they soon figured out what they needed to do to accommodate the lumpiness of space.

We scientists can be real worriers sometimes. We see something new in our data and our scientific training teaches us to be suspicious. Perhaps, we think, the instrument has had a hiccup. Perhaps there has been a little spark in one of the other instruments that has affected our gizmo. We seek independent evidence to confirm that what we have seen is real. If we cannot find that evidence, the result is consigned to the back of our minds as something strange that we might understand some day. Until then, it is not for public view.

Looking back on *Pioneer 6* and 7, I can see that our scientific training may have caused us to miss a very important discovery. On rare occasions we would see what appeared to be a great burst of radiation, quite unlike anything we had ever seen from the Sun. The bursts were enormous and would send our very sensitive detector way off the scale. They would last for a minute or less, compared with many hours for the bursts from the Sun. We considered all the possibilities we could think of, to no avail. We could not find out what was producing these strange bursts. There were other things to study in the data, there were other spacecraft instruments to build, so we forgot about these strange bursts.

At the time there were several military satellites orbiting the Earth to warn of rocket launches anywhere on Earth. Using data from these, the scientific staff noticed several big bursts of radiation similar to those we had seen. They also noticed something strange about them: two spacecraft would see a burst at approximately, not exactly, the same time, perhaps several seconds apart.

Comparing the results from two spacecraft—one near Earth and the other a long way away—the difference was a minute or more. They realised that the bursts were caused by gamma rays reaching the solar system from other places in the universe. They are now believed to be the result of the collapse of a star in another galaxy and known as 'gamma-ray bursters'.

It's just one of the ones that got away. Perhaps my colleagues and I might have been able to make that discovery if we had looked hard enough, but that's the fun and challenge of science. Sometimes you miss something and someone else finds it; at other times you might make an important discovery that someone else has missed. *C'est la vie.*

‹ 24 ›

Give them a shovel

I received the phone call two years later, asking me to serve on a committee whose job it was to decide what should be done to stop the American astronauts from being killed or losing their fertility on their way to the Moon.

Several years after arriving in America, I received a phone call asking me whether I would act as a consultant on the American manned space program. There was a concern that the astronauts could be killed by an intense burst of solar protons thrown out by a large solar explosion.

There was real cause for concern. In 1956 the Sun had produced a massive explosion, which had sent radiation counters all over the world off the dial. (See Chapter 7 and Figure 1)

At some points on Earth, the radiation was 40 times stronger than normal. Soon after a rumour started to circulate within the scientific community. It said that the aircrew of a number of the B52 bombers of America's Strategic Air Command (SAC) had received very severe doses of radiation during this radiation burst.

It was a rather plausible rumour. The cold war was in full swing, and the Americans were afraid that their bombers would all be caught on the ground in a pre-emptive strike like Pearl Harbour, except with rockets and hydrogen bombs.

> *A rumour started to circulate ... that the aircrew of a number of B52 bombers of America's Strategic Air Command had received very severe doses of radiation during this radiation burst.*

To the military mind, the solution was simple: keep the bombers in the air as America's nuclear deterrent. SAC bombers would fly round and round for hours, prepared to retaliate immediately should any hostilities commence. To be as near to their targets in Russia as possible, many of them would circle over the Arctic cap, which was exactly where intense radiation from the Sun would be focused by the Earth's magnetic field.

Having heard the rumour and having access to the data recorded at the time, I calculated the radiation dose at an altitude of 50,000 feet (15.2 kilometres).

The results of the calculation were not cheerful news. I worked out that someone, somewhere, should be quite worried. Perhaps around 100 airmen, for example, might have been concerned for their virility. The radiation in question has disastrous effects on the reproductive regions of the human body, causing the testicles to forget what they are supposed to do or causing serious damage to a fetus in the uterus.

The first consequence of this calculation was immediate. My wife and I were expecting our second child. We would soon be flying back to Australia, possibly by way of Japan. There was *no way* that we would fly across the Arctic region to Japan.

Other things started to happen. NASA started to become edgy about something. They commissioned me to calculate the characteristics of the solar cosmic radiation. For ten days work they paid me 10 per cent of the price of the fancy four-bedroom house we were buying.

I also began to hear of studies in Britain. They were doing something I had done several years before, when I proved that there was one region on the Sun from which radiation would reach the Earth with little reduction in intensity.

At the time the British and their great friends the French, were building the world's first supersonic passenger aircraft, *Concorde*, which would fly at 70,000 feet (21.3 kilometres), exactly where the effects of the radiation would be worst. (Again Murphy's Law was coming into effect.) Soon we began to hear from our British colleagues that they were designing a radiation detector to go in *Concorde*.

Consequently there was no doubt in my mind that when President Kennedy announced that America was going to the Moon, he was also announcing that there would be good jobs for cosmic ray physicists who understood the inner workings of the solar system.

It was no great surprise when I received the phone call two years later, asking me to serve on a committee whose job it was to decide what should be done to stop the American astronauts from being killed or losing their fertility on their way to the Moon.

On the appointed day I arrived at the Johnson Space Center, in Houston, Texas. Together with the other members of the committee I learned that the powers-that-be had decreed that the *Apollo* mission to the Moon was to have a 99.5 per cent probability of success or, to put it another way, odds of 200 to 1. These figures trip off the tongue easily and sound very comforting. Since there were about a million different things that could go wrong, it meant that we had to reduce the odds of any single thing going wrong to about 200 million to one.

Since there were about a million different things that could go wrong, it meant that we had to reduce the odds of any single thing going wrong to about 200 million to one.

The immediate reaction was: 'You're joking!' However, we learned very quickly that the manned space flight people were pretty uptight types and didn't joke about anything. They meant exactly what they said and we had the clear impression that, if they had their way, they would fly 200 million flights to the Moon to make certain we had done our job properly.

So we started to decide what precautions would be necessary to reduce the odds of the astronauts being zapped to one in 200 million. To help us to understand the problem, we were taken to the prototypes of the *Apollo* spacecraft and the 'lunar excursion module' (the LEM) and allowed to crawl around inside them. We measured things, and asked the engineers to tell us about the manoeuvres that the spacecraft and the LEM could perform, if required. To be honest, it was fun.

And so to work. We learned that very little was know about the effects of radiation on the human body. Most of what was know derived from studies of the survivors of the atomic bombs dropped on Nagasaki and Hiroshima in 1945. For the worst radiation burst known till then (our friend from 1956), calculations indicated that one in three people would have exhibited the first signs of serious radiation sickness—if they had been on the Moon, outside the protective shield of Earth's magnetic field. Anything bigger would be extremely serious.

We considered the construction of the *Apollo* spacecraft and decided that if it were hit with a similar burst the astronauts were relatively safe, as long as they stayed in the spacecraft. Its relatively thick walls would reduce the cosmic radiation to a tolerable level.

The problems started when they moved into the LEM, which was going to take them down to the surface of the Moon, or when they were on the surface itself. The LEM was built out of aluminium sheet about the thickness of cooking foil, and it would give them absolutely no protection.

Of course, no one had been to the Moon yet either. We had no idea what the surface would be like. One of my friends, a rather wild theoretician called Tommy Gold, had predicted that the surface would be covered by a 20-metre (65 feet) thick layer of talcum powder. Others thought it would be rock hard. Others thought the consistency of fine sand. We simply did not know.

We talked. We thought. We argued. There was absolutely no way in which the LEM could be changed to give the astronauts any reasonable protection. The rockets that would take the LEM back to the *Apollo* mother ship did not have enough oomph to add any more shielding.

Our conclusions? Firstly, choose the time to go to the Moon when there were no big sunspots on the face of the Sun. Secondly, do not leave the spacecraft if the Sun had, perversely, suddenly produced a big sunspot while *Apollo* was on its way.

The NASA engineer who convened the committee was not satisfied. He asked, 'But what about when they're on the Moon', and kept reminding us that the astronauts would be on the surface for more than a day.

He persisted, 'What if one of these big bombs goes off on the Sun when they are there—what do they do?'

At last the committee gave him an answer—he didn't like it, but that was his problem.

The answer was to equip the LEM with a shovel and make certain that the astronauts could dig holes very quickly. We told him they would need to dig a hole at least three metres (10 feet) deep in an hour to give them any protection.

> *The answer was to equip the LEM with a shovel and make certain that the astronauts were extremely good at digging holes very quickly.*

NASA carefully acted on the advice we gave them about avoiding the radiation problem by not going to the Moon while there were large sunspots active on the Sun's surface and staying in the *Apollo* spacecraft if one appeared while they were in space.

They decided not to send a shovel.

Some years later the author James Michener wrote *Space*, his blockbusting book about the American space program. It was fiction, draped loosely over the basic facts of the *Apollo* program.

To those who had been involved in the space program, it was an interesting game to 'spot the real life person' who appeared as a fictional character.

You may recall that *Apollo 17* was the last Moon mission by the Americans. Well up to that point in his novel, Michener pretty much kept to the historical facts. Then he launched *Apollo 18* and a young astronomer, who is a consultant to NASA, explains his discoveries about how cosmic radiation travels from the Sun to the Earth.

'Hang on a moment,' I thought, 'those are my discoveries. Clearly I have a walk on part in the book'.

Things start to go seriously wrong at that point in the book. An enormous sunspot appears. The astronomer does not see it and *Apollo* is launched. The astronauts go to the Moon in the LEM.

KABOOM! An massive explosion occurs on the Sun and all the worst things our committee predicted happen. The astronauts don't have a shovel and they die within an hour.

I was rather chuffed that I, and our committee, appeared in Michener's book, but perhaps I boasted about it once too often.

One day at dinner our 18-year-old son said, 'Dad, I've gone back to read about you in Michener's book. I think you have some explaining to do?'

Muggins here didn't sense the danger. 'Stuart, how come?'

'Well, the book explains that the astronauts were killed because you didn't see the big sunspot. The reason you didn't see the big sunspot was that you were having it off almost continuously with your girlfriend. That would have been 1972 when I was six years old. Dad, do you have some serious explaining to do to Mum and us kids?'

Sometimes kids are too damned smart.

‹ 25 ›

Someone has to do it

There were 53 proposals. They were all mailed to us, all 500 pages or so, about a week before the committee met. Then we met for two days at NASA headquarters in downtown Washington, DC.

Peer review: it's the process used by government agencies to decide who will receive funding for their scientific research. It's used by publishers to decide whether a book is good enough to be published. Scientific journals use it to decide whether the results of research are worth publishing. Since scientific research is usually about exploring the unknown or testing controversial ideas, and since the allocations of very large sums of money may be involved, decisions such as these can arouse strong passions.

To make the process as objective and defensible as possible, the decisions are usually made by a committee of well-regarded scientists who are experts in that field.

Following the flight of *Sputnik 1*, NASA was provided with large amounts of money to support space research. There was money for rockets and satellites, money for new professors and graduate students, money to build and equip new space research laboratories at many American universities.

Soon one of the most important parts of a professor's résumé was their success in securing research funding. The most effective way to win large amounts of money was to be given a ride on a NASA satellite. So, once the rockets had become reliable, there was intense competition for every satellite NASA decided to fly.

However, this was still the 1960s and there weren't many experienced space scientists about. Some of the experimenters on the first satellites had decided not to risk their sanity any further and left the field. There were a few with real experience, like Jim van Allen of radiation belts fame. Then there were some, like me, who would have experiments in space in the near future and had a good idea of the science and the agony involved. This didn't give NASA a huge choice for its peer review panels and so it was that I was asked to serve on the 'particles and fields subcommittee' in 1964.

At the tender age of 30, I was somehow deemed an expert in space research. I was also a foreigner who had already overstayed his two-year visa by three years. Together with the five other members of the committee, I would make recommendations that would lead to the expenditure of hundreds of millions of dollars. The future and prestige of American space research could be irrevocably affected by our recommendations. Whichever way you looked at it, this was a surprising outcome for someone who had been a complete unknown in international science just five years previously.

I'll describe the selection process for an early 'observatory' class satellite, which was planned to be flown at the end of the 1960s. It was to orbit the Earth, ranging from a low altitude of about 300 kilometres, to a high point about a third of the way to the Moon. Allowing for the available space in the satellite and the weight, power and data transmission capability, we were asked to select seven candidate instruments for flight. They were to be in order of priority, so one or two could be excluded later without discussion.

Each aspiring experimenter submitted a proposal, describing the instrument in ten pages or less. This was to include a careful discussion of the prevailing scientific knowledge and how the proposed observations would contribute to that. The power required, the weight and, perhaps the most important, how much data needed to be transmitted to Earth, were vital numbers as well.

There were 53 proposals. They were all mailed to us, all 500 pages or so, about a week before the committee met for two days at NASA headquarters in downtown Washington, DC.

We started by separating the experiments into classes: magnetic field measuring instruments, instruments to measure the electrons in the van Allen belts, instruments to measure the solar wind, the cosmic rays and so on. Then the hard part began, choosing between them.

For example, there were eight proposed experiments to measure the electrons in the van Allen belts. Some were little more than copies of the instruments used by van Allen in 1957. Some were very complex and some of us would argue whether they would ever work. Some were very ingenious, and were discussed with admiration by the members of the committee.

Then each of the proposers was asked into the meeting to give a five-minute summary of the experiment and its importance to the American space program. The presentation was followed by five minutes of robust questioning by the committee.

Having heard the presentations for that class of experiment, the committee would argue among itself for an hour or so about the relative merits and demerits of each. Finally the chairman said, 'That's enough talk, it's decision time'.

Usually about two thirds of the proposals could be eliminated immediately. Then we needed to make an agonising choice between two or perhaps three very good proposals.

It was never easy. Frequently the committee would backtrack to an earlier decision, often stimulated by discussion of another instrument. Inevitably the committee ran out of time. About a half hour before we needed to run off to the airport, the chairman would call for a sudden-death decision. There would then be an argument about which of the 'winners' were of lower priority.

We could never agree. There would be a vote. Sometimes the chairman needed to make a casting vote. The reality was that only five years into the space era, there were more good experiments than could be accommodated within NASA's commodious budget.

It was going to grow much, much worse.

Year by year more space scientists achieved their doctorates. A trickle became a torrent. More and more were employed by universities in the belief that this would attract large amounts of research money—40 per cent of which went immediately to the administration of the university. Many more experiments were proposed, increasingly from universities and research institutions that were new to space research.

By the time *Apollo 11* went to the Moon in 1969, the space budget had stopped growing. Already there was a backlog of satellites that had been approved, experiments chosen, but not funded. NASA said that they would honour those commitments. As a consequence, there was very little new space funding in 1970. The space community was starting to hurt badly.

By then the space shuttle was on the drawing boards and NASA began planning for 'laboratories in the sky'. As a foretaste of this new post-Moon era, NASA announced that they would use a rocket casing left over from the Moon program to build Skylab.

It would be a 13-metre-long (43 feet) cylindrical tube, four metres (13 feet) in diameter and would allow up to seven scientists to perform their research in a shirt-sleeves environment—that is, without a space-suit. There was a great deal of ballyhoo about this being the 'first manned base in space' devoted to scientific research and so on. To match the ballyhoo, $50 million was removed from other space projects and earmarked for scientific instruments on Skylab. As was their wont, NASA called for proposals from the American scientific community.

As I have outlined, people were hurting in NASA laboratories, universities, everywhere. So the obvious happened; everyone proposed to fly something. This posed a big problem. NASA had become sensitive to criticism that the selection process had become an 'old boys club' and had decided to become squeaky clean. There would be no conflicts of interest they said. If there were a proposal from someone at MIT, for instance, then no one from MIT could be on the selection panel.

As a consequence, there were no suitable Americans left to serve on the selection committee. Consequently NASA roped in a couple of Poms, a Frenchman, a German, a Japanese and yours truly from the Southern Hemisphere. At great expense, they flew us from the ends of the Earth, first to San Francisco then, some months later, to Washington, DC to sit in judgment over the American space research community.

Murphy's law struck again. Our three-day meeting in Washington coincided with massive marches against the Vietnam War. Traffic was at a complete standstill, making it difficult for the committee to attend the meetings. More than a hundred people associated with the proposals were delayed as well. It threw the meeting into chaos.

There were many excellent proposals. Some were extremely well suited to *Skylab*, and went on to make important discoveries. There were some that exercised our minds a great deal, for both practical and scientific reasons.

An internationally recognised scientist presented one memorable experiment to detect anti-matter protons in the cosmic radiation. Anti-matter was then a seductive subject for physicists in general. We knew the world around us was made from atoms in which the nucleus carried a positive electrical charge, surrounded by electrons that were negatively charged.

The theoreticians, however, had asked: 'Why can't we have atoms with a negatively charged nucleus and positively charged electrons orbiting about it?' This was known as anti-matter.

We had already seen positively and negatively charged electrons in the atmosphere and observed that, when they met, they mutually destructed, giving out a burst of 'annihilation radiation'. So where were the anti-matter protons?

To find anti-matter, the scientist proposed to use a large 'cryogenic' magnet, which would generate very strong magnetic fields to deflect the 'matter' protons in one direction and the 'anti-matter' protons in the opposite direction. This was an elegant idea, and we liked it. However, there was a practical problem.

Cryogenic magnets only operate if immersed in liquid helium. Over time the helium evaporates and eventually the magnet will warm up. Magnetic fields contain a lot of energy, so the question was: 'Will the experiment explode in an uncontrollable manner?'

There had been explosions already in a number of laboratories on Earth. An explosion on board a thin-walled space laboratory didn't seem like a good idea.

The speaker knew we were worried about this problem. He gave some half-baked assurances in his five-minute presentation, but could see that we weren't buying it. In some degree of frustration he said, 'Look, I think you are worrying about this too much. You are exaggerating the problem. It's like when you go to a conference and the person you sleep with … er, the person who shares your hotel room, says that you snore, when you only make the occasional noise in your sleep. I am certain that the investigations we are making right now will show that it will be perfectly safe'.

His presentation finished, our German chairman invited us to ask questions. Quick as a flash, one of the British committee members leaned across the table and said, 'Hey mate, regarding your suggestion that we are exaggerating the cryogenic problem, tell us something more about the people you said you sleep with'. The German chairman was somewhat lacking in a sense of humour and could not quite understand why it took about five minutes to put the meeting back on track.

'Regarding your suggestion that we are exaggerating the cryogenic problem, tell us something more about the people you said you sleep with'.

The meeting continued and we reached an experiment proposed by Dr Wernher von Braun. The committee blinked collectively. Things must be very difficult out there, we thought.

In an earlier chapter I described how Germany, in the last desperate days of World War II, sent the *V2* rocket to terrorise Britain. The leader of the German rocket program was Wernher von Braun. At the end of the war there was a frantic rush by the Russians and Americans to 'liberate' Wernher and his mates. The Americans more or less won. They claimed Wernher von Braun, along with about 100 unused *V2*s. The Russians took other members of his team. From these sprang the Soviet and American space programs.

Von Braun then went onto bigger and better things. While the American military fired off the liberated *V2*s in the deserts of New Mexico, von Braun persuaded the military and politicians to build rockets big enough to dump hydrogen bombs on Moscow. By the late 1950s he had well and truly succeeded.

Wernher needed a new challenge. He wanted to build even bigger rockets. To his great relief, America decided that it wanted to put men the Moon. So he and his acolytes started to build *Saturn 5*, a 96-metre (315 feet) high behemoth, which nine years later took Armstrong and Aldrin to the Moon. All this made Dr von Braun very famous, or infamous, depending on your point of view.

There was a mathematics tutor at Harvard University in the early 1960s called Tom Lehrer who moonlighted as a folk singer. He did a good trade in rather clever songs about the periodic table, mathematics to base eight and the environment. His song about the environment: 'Fish gotta swim and birds gotta fly; but they don't last long if they try', is still as relevant as it ever was.

Lehrer was rather popular with the scientific crowd. He did no harm to his reputation in Australia when he was officially banned during a visit in 1961. A song about the boy scout movement and a somewhat sexual version of their 'be prepared' motto aroused the censors from their torpor and he was told to go. Probably the first scientist to achieve that distinction.

I digress. Lehrer satirised many things. Politicians and social and scientific hypocrisy were favourite targets. Wernher von Braun was one of his targets. Those of my generation of scientists will never forget the line:

'"Ven zee rocket goes up, who cares vhere it comes down, that's not my department", said Wernher von Braun'.

Back to our committee. So here we had Dr von Braun wanting his piece of the *Skylab* money. Things must have been tough out there since von Braun seemed to be short of the odd $10 million or so.

As with every other proposer, von Braun was allowed five minutes to emphasise the main reasons why his experiment should be flown ahead of all the other candidates, followed by questions from the committee. Von Braun's turn came. The chairman went out to fetch him. The committee blinked collectively again. Not just one or two people came in—there were ten.

From the beginning, it was clear that von Braun felt that this was all below his dignity. He, who had dealt with presidents, Führers and four star generals, had to beg money from these insignificant scientists.

To make certain we understood this, he actually stated that this was all below his dignity and that it was a poor reflection on the American character that he now had to beg money from insignificant scientists; foreigners yet. Having put us at ease in this manner, he handed over to one of his lieutenants to explain what they wanted to do. They wanted three times as much money as was available, in fact between ten and fifty times what anyone else asked for.

To make certain we understood this, he actually stated that this was all below his dignity ...

As I recall, none of the committee wanted to ask any questions. Von Braun was clearly of the view that that was right and proper, and that our lack of questions was an indication of our high regard for his proposal.

Later the committee made its recommendations to NASA. Many excellent experiments were flown and significant discoveries were made. Dr von Braun's experiment was not one of them.

‹ 26 ›
Woomera

Having obtained official access to Woomera, all the technical people in Australia were extremely helpful. The engineers and technicians at Woomera and its supporting laboratories in Elizabeth, near Adelaide, were nearly all Australians. They had long fulminated against the fact that no Australian experimenters were using the Australian half of the Woomera facility.

In 1965 I was invited back to Australia to be interviewed for the position of Professor of Experimental Physics at the University of Adelaide. The Woomera rocket range was some 400 kilometres (250 miles) north of the city and it seemed to me that this fact, combined with my NASA contracts and other activities, would allow me to continue space research in Australia.

Starting in the 1940s, high-altitude sounding rockets were used to measure the properties of the Earth's atmosphere to heights of 50 kilometres (30 miles) and more. Like a cricket ball, they would leave Earth at a high speed. Gravity would gradually slow down their ascent, they would ultimately attain their maximum altitude, then fall back to Earth. They would only spend minutes at high altitudes, but at a time when we knew little about the atmosphere and space beyond, this was quite enough. We were thankful for whatever information we gained.

Woomera was a fantastic rocket range. Rockets were launched from near Woomera township and fired in a north-westerly direction to land in a long, thin region of the desert, an area of some 1600 square kilometres (620 square miles)—all a long way from the nearest sea.

So while at most of the other ranges in the world rockets would fall into the sea and be lost, at Woomera you could retrieve your scientific equipment. Furthermore, it was well equipped and the only serious rocket range in the southern hemisphere.

As a consequence rockets flown from Woomera would allow us to study almost 50 per cent of our galaxy—little of which could be seen from the rocket ranges in America and Europe. Finally, probably the sweetest, most beautiful rocket ever flown, the British *Skylark*, was flown from there about twelve times a year—often almost empty.

Officially, Woomera was a 50–50 joint project between Australia and the United Kingdom. It was run by the Weapons Research Establishment (WRE), an arm of the Australian Department of Defence. I knew, however, that it was used for civilian research, and that lots of scientists from British universities were flying their experiments on rockets flown from Woomera.

To my mind, since it was sold to the Australian taxpayer as being 50 per cent ours, I figured that we Australian scientists should also be given a go.

This was not, of course, an original idea. Australian scientists had tried to secure rides from Woomera during the International Geophysical Year of 1957–58. As a result of their approaches, the

Australian government set up a committee of scientists to look into it. Years went by … and it was decided that it was too important for scientists to look into. They needed a committee of the most senior university people possible—the vice chancellors.

Another committee was set up. Now, vice chancellors are busy people. More time went by. Finally a report was published, which made a good case that, since many of the best rockets were being flown almost empty, it would be a wise use of resources and would assist scientific research and education if the scientists could hitchhike on these rockets.

There had been other approaches, too. Following the launch of the *Sputnik* satellites, America had decided to encourage other anti-communist countries to become active in space research. To this end they offered large rockets, and all the support needed to launch them, to a number of their European friends. The offer was made to Britain, France, Germany and Italy, and each rapidly accepted and flew their own national satellites over the next few years.

At that time an Australian, Dr David Martyn, was chairman of a United Nations committee, which had been established to consider the scientific issues associated with the 'peaceful use of outer space'. This led to Australia being offered a rocket as well. An inter-departmental dispute within Australian bureaucracy led to this offer being turned down.

However, back to the vice-chancellors' report. The wheels of government grind slowly. So it was that, seven years after the first request before the International Geophysical Year and in view of the good arguments made by the vice chancellors, the answer was, 'No'. The prime minister of the day said that Australia was too small a country to want to go to the Moon.

The prime minister … said that Australia was too small a country to want to go to the Moon.

Actually, no one had ever said anything about going to the Moon—250 kilometres (155 miles) into space was far enough for the needs of Australian science and industry—but that made no difference. That was the government's answer. That it was the answer to a question that had not been asked was conveniently overlooked.

Mercifully, I had not been involved in this earlier, frustrating collective bashing of heads against the assorted pricks of ill fortune in Canberra. I had been in America, merrily flying big balloons and building satellites. So I was fresh to the fray.

Having been interviewed for the University of Adelaide position, I arranged to visit the director of WRE. On being ushered into his room, he said, 'Oh, McCracken, I've heard of you. Let me make it clear from the outset that there is no way that you can get access to the rocket range or our facilities'.

Hoping to soften his attitude somewhat, I talked about our *Pioneer 6* and 7 spacecraft and the *Interplanetary Monitoring Spacecraft* we were designing. I told him about our discovery of an intense X-ray star using the balloon flown from Hyderabad in India. I spoke of the important role that Woomera could play in this newly discovered field of X-ray astronomy. He was totally disinterested. I asked whether I could see the test facilities that were used to shake, cool and heat the rocket payloads prior to flight. He said that this could not be arranged. Clearly, this was not the way to go.

Perhaps I could have gone to Canberra and pointed out what I had done in America, and the resulting industrial payoff. Smaller countries, including the Netherlands and Sweden, had initiated space programs to allow their aviation, electronics and manufacturing industries to compete in the new 'space age'. They incorporated the resulting 'spin offs' throughout their economies.

We were ... drinking English cider, which I find has a marvellous ability to allow one to think without inhibition. Having had enough cider to ignore all reasonable prudence, I thought of a crazy idea.

I could have argued that it would be in the national interest for Australia to have a modest space program. Then there would have been another committee, which would have put the noses of the scientists from the earlier committee out of joint. Another seven years might have gone by.

No, that was not the way to go.

Six months later I was at an international conference in London and discussing the problem of how to gain access to the rockets flown from Woomera with my old boss, Geoff Fenton, from the University of Tasmania. We were in the staff club of one of the London universities, drinking English cider, which I find has a marvellous ability to allow one to think without inhibition. Having had enough cider to ignore all reasonable prudence, I thought of a crazy idea.

Plate 1: The young Ken McCracken (aka 'Lic lic skule boi') and his 'wash-boi' in Papua New Guinea, 1957.

Plate 2a: The author's cosmic ray observatory on the slopes of Mount. Wellington, above Hobart, Tasmania, 1956.

Plate 2b: The author performing the weekly check on the operation of the neutron monitor in the cosmic ray observatory, 1956.

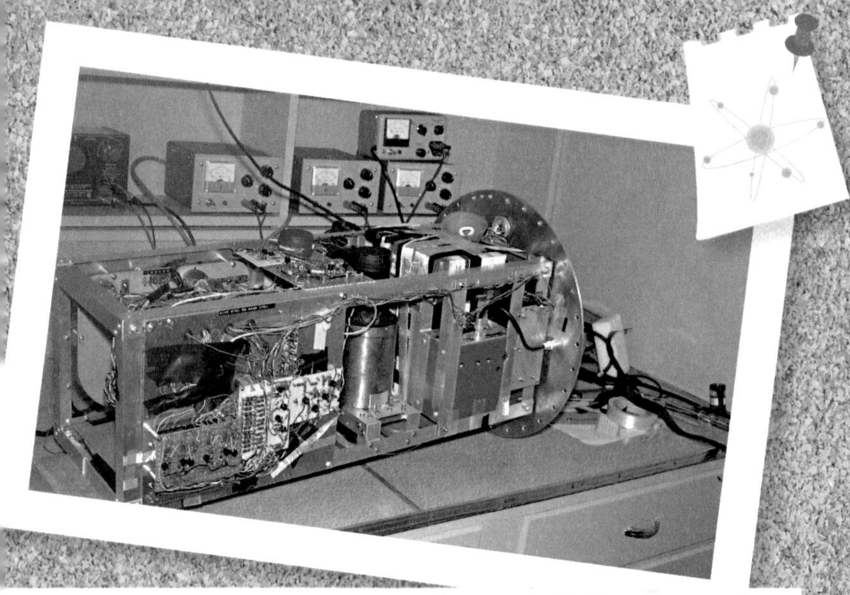

Plate 3a and 3b: A self-contained balloon payload (above) to test the author's ideas for the Pioneer *spacecraft. This contains the cosmic ray instrument, a radio transmitter and tape recorder, all the batteries and other test instruments needed to interpret the data. It was then surrounded by a large amount of thermal insulating material and hung below the large, helium-filled balloon (below) that swelled to a diameter of 100 metres (330 ft) when it reached an altitude of 32 kilometres (20 miles).*

Plate 4a: The cosmic ray anisotropy instrument flown on the Pioneer 6 *spacecraft.*

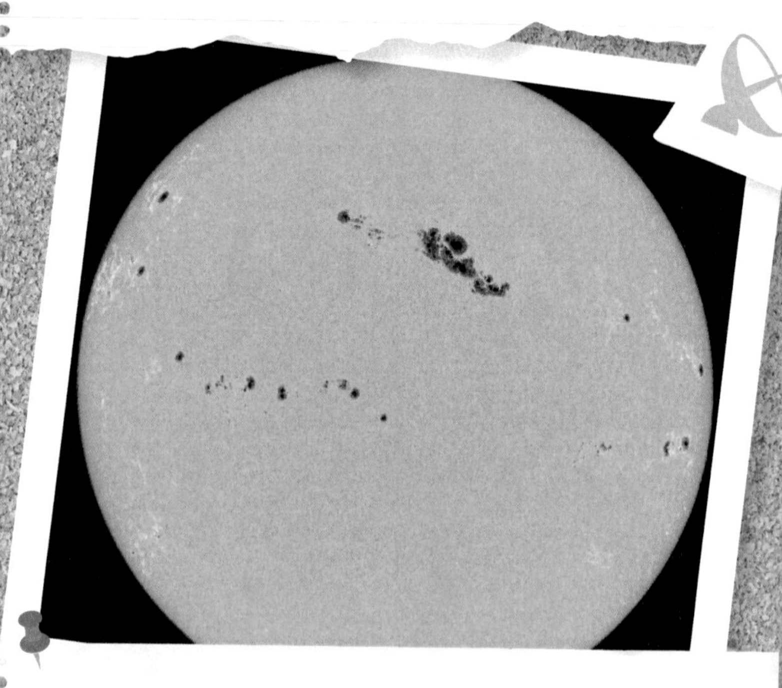

Plate 5a: The Sun, with two large groups of sunspots. (NASA)

Plate 4b: The author's improved anisotropy instrument flown on Pioneers *8 and 9. It gave more accurate directional information, and detected cosmic rays over a wider energy range than the earlier instrument.*

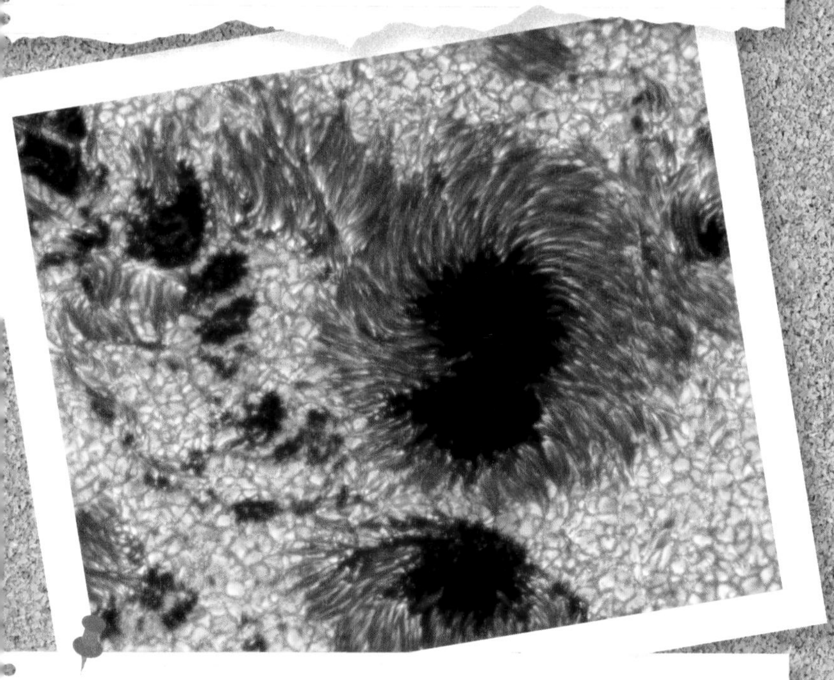

Plate 5b: A very high-resolution image of a medium-size sunspot group, taken in July 2002. The larger spot has a diameter of about the size of Earth. (Royal Swedish Academy of Sciences).

Plates 6a and 6b: The Skylark *X-ray astronomy experiment that discovered the extremely bright X-ray star near the southern cross. The 'venetian blinds' restrict the vision of the detector to a small portion of the sky.*

Plate 7: Ken McCracken (right) with his mentor and advocate, Professor Bruno Rossi, of the Massachusetts Institute of Technology.

Plate 8: A Skylark *atmospheric 'sounding' rocket launch from Woomera, South Australia (DSTO).*

The next day Geoff and I took the underground to the British National Space Centre—the British equivalent of NASA. We met with their senior officers and explained that we wanted to look for X-ray stars in the southern sky, preferably using several rockets flown from Woomera

I then pulled my passport from my pocket. In those days we Australians carried British passports. I explained that we wanted to be allowed to supply experiments, at our cost, for the British *Skylarks* being flown from Woomera. I stressed that we wanted to be regarded as British, not Australians.

The British, after recovering from their laughter, were very encouraging. They told us to write a letter explaining what we proposed and they would consider it.

We did, the next day. They did consider it, quickly. So it was that within a week, those well-known British institutions, the Universities of Adelaide and Tasmania, were officially accepted as part of the British space program at Woomera. We were assigned space on two *Skylarks*, SL425 and SL426, to be flown in early 1967.

The great virtue of being the first to make any new scientific measurement is that the experiments can be, and even should be, very simple. So our first two X-ray astronomy payloads were remarkably easy to build. The British National Space Centre supplied us with the outer skin of our section of the rocket and some simple design restrictions to make certain everything would not fall apart. The X-ray detectors were purchased in America using my NASA funds and everything else was built in Hobart and Adelaide.

Having obtained official access to Woomera, all the technical people in Australia were extremely helpful. The engineers and technicians at Woomera and its supporting laboratories in Elizabeth, near Adelaide, were nearly all Australians. They had long fulminated against the fact that no Australian experimenters were using the Australian half of the Woomera facility.

Once they were allowed to assist us, we were showered with an embarrassing amount of help. In particular, we were immediately asked if we would like to use the six almost empty *Skylarks*, which were flown each year as the Australian share of the flights at Woomera. Would we ever! To our great relief, the Australian technical people were determined to see that the locals were given a fair go.

The British were equally helpful. Each week they flew a Comet 4 jet airliner from the United Kingdom and back to transport their staff to the rocket range. We were immediately invited to use these flights free of charge if there were vacancies, which there almost always were. This allowed us to sort out technical problems in the UK before the rockets were sent to Australia. In the early 1970s, airfares were much more expensive than now, in real terms, and this was a most welcome and practical help.

It occurred to me, however, that Canberra might not be entirely happy that we had gained access to Woomera over the back fence, as it were. We therefore kept very quiet about what we were doing. No reports to conferences. No press releases. Definitely nothing in any report that went within a bull's roar of Canberra.

There was the ever-present fear that officialdom in Canberra would find a way to stop us if it caught wind of our activities.

The great day of our first flight fiinally came. Woomera is a long way out in the desert and the only accommodation was in a small township run by the rocket range itself. They could only provide accommodation for two people so I sent the two graduate students who had built the X-ray telescope. The rest of the team sweated it out in Adelaide and Hobart. The launch was scheduled for 2.30 pm and we had instructed the students on the ground at Woomera to telephone us immediately it took place to tell us what had happened.

Soon after 2.30 pm my senior technician ran up to my room and told me that the ABC had just released a 'newsflash' stating that the cabinet minister responsible for Woomera had announced the launch of a rocket as part of the Australian space program to search for X-ray stars in the southern skies.

With this I learned a truth that comes to all young scientists in the end. Sometimes officialdom will not assist you one iota, but when you succeed in spite of everything they do, they will claim the credit.

Sometimes officialdom will not assist you one iota, but when you succeed in spite of everything they do, they will claim the credit.

I immediately wrote to the cabinet minister who had claimed the credit for what we had done. I pointed out that given support—that is, money—we could do a lot more good science and assist Australian industry as well. I received a non-committal response, which I later learned was typical of the bureaucrat: lots of platitudes and no assistance.

The Brits had given us two rides and our second rocket was launched just two weeks after the first. We analysed our results, and our first flight showed strong X-ray emissions from stars in the southern sky. One of these, at the centre of our galaxy, had been seen from the northern hemisphere. This gave us confidence that our detectors were working properly.

In those days rockets would spin and tumble in space and it was not easy to know which way our X-ray telescopes were looking. We had worked out a method to do so and that bright source in the centre of the galaxy proved that our mathematics were right. However, the others sources of X-rays were further south, invisible from the northern hemisphere. We were exploring terra incognito, so to speak. One of these new X-ray stars was very bright and we became very excited. We were onto a big discovery.

During the previous year or so we had suggested, in jest, that we were searching for an X-ray star in the Southern Cross. My students and I worked out the position of each of the new X-ray stars we had seen and found that our very bright new X-ray star actually was in the constellation Cruxis—the Southern Cross!

Panic. We had never been serious about our suggestion and now the damned thing *was* there. Had we fallen into the trap of the self-fulfilling wish?

With great care we checked every calculation and every assumption. No matter what we did, the X-ray star stayed stubbornly fixed in Cruxis. 'So', I said, 'let's quickly confirm its position with the data from the second flight', which had occurred just days before.

Then we had our second big shock. The X-ray star in the centre of the galaxy was shining brightly, unchanged, but our bright X-ray star in the Southern Cross was now less bright than two weeks previously.

How could this be? The theoreticians, having recovered from their first assertion that X-ray stars were impossible, were now equally adamant that they should not be able vary in intensity with time.

We experimental physicists are a stubborn tribe, however. Often we have a pathological distrust of theoreticians, so we checked and rechecked both sets of results. We eventually convinced ourselves that we could trust the data and our interpretations.

So just one week after our second flight, I announced our discovery of a new X-ray star in the southern sky whose strength fluctuated with the passage of time. The announcement was made in Washington, DC, at the most important space research conference that occurred that year. I detailed the observations, and the way we had verified the accuracy of our measurements using the observations from the northern hemisphere. I stressed that the two flights gave unambiguous proof that X-ray stars could vary in intensity very quickly.

In the audience were two of the top American scientists in X-ray astronomy. It is on record that one leaned over to the other and said, 'Walter, do you believe this rubbish?' Twenty-five years later Walter Lewin published an account of the early days of X-ray astronomy. He identified our paper at that conference—and the question he was asked—as one of the most exciting points in the development of this entirely new branch of astronomy. X-ray stars are now known to be very fickle, waxing and waning with a promiscuity unknown anywhere else in astronomy. It was our 'unofficial' flights that discovered this.

Several years later my colleague in the physics department at the University of Adelaide, Professor John Carver, received an unusual offer. Would he like to fly instruments on a satellite to be launched from Woomera using a large rocket left over from a series of flights America had been making?

The Americans had been flying *Redstone* rockets from Woomera to determine how hot their nose-cones became as they re-entered the atmosphere from space at very high speeds. These were much larger rockets than the small *Skylarks* we had been flying with such success.

The *Redstone* was, in fact, a souped-up version of the *V2*. It was also the type of rocket that was used to launch the first American astronaut into space. This was one very serious rocket.

There was a catch, however. The American investigations would be ending soon, and the agreement with Australia was that America would take any spare rockets back with them. Launching one was, in fact, the simplest way to overcome this problem. Giving it to Australia was even more of a win–win situation.

The catch was that the satellite had to be designed, all the bits bought, the satellite built, tested and successfully launched, all in ten months. This was a terrifying prospect. Frequently the delivery time for space components was a year or more. Many corners would need to be cut.

Nevertheless, the offer of a free rocket was accepted. The satellite would be called *WRESAT*, standing for Weapons Research Establishment Satellite. It was to be an 'in-house' program, conducted in the facilities of the WRE in Elizabeth, South Australia. The only outsiders were John Carver and his students from the University of Adelaide. They had been studying the absorption of ultraviolet light in the atmosphere using small rockets and their detectors were almost ready to go.

It was touch and go. Several days before *WRESAT* was due to go to Woomera, it was still malfunctioning in a very bad way. I was asked to attend a meeting to think of ways to work around the problem. The project manager stated firmly that if it wasn't working at the time of launch—too bad! 'The launch must proceed anyhow.'

The gods were kind. The problem was identified and rectified. *WRESAT* roared into orbit on 29 November 1967. The short time available to build *WRESAT* meant that solar cells could not be obtained in time. Consequently it was powered with batteries that ran out after a month in orbit. Nevertheless, it provided excellent data and made important contributions to our knowledge of the Earth's upper atmosphere.

John Carver had hoped that *WRESAT* would be the first of a series of Australian satellites. I had hoped that we could continue to use *Skylarks* for X-ray astronomy.

Unfortunately, neither eventuated and space research in Australia became entirely dependent on 'getting rides' on satellites and rockets flown by other nations.

One such Australian project was the Orbiting Satellite Carrying Amateur Radio, built by a group of radio amateurs at the University of Melbourne and known as *Oscar-5*. It roared into orbit in 1970 in a piggyback launch with an American meteorological satellite.

Thirty years later the Cooperative Research Centre for Satellite Systems (CRC) was established by the CSIRO, several universities and five industrial companies.

Funds were provided by the Australian government to build a scientific research satellite to celebrate the centenary of the Federation of the Australian colonies in 1901. It was named *Fedsat* and on 14 December 2002 it carried into orbit a number of experiments built by Australian universities.

The Director of the CRC, Brian Embleton, and his university colleagues achieved what my generation had failed to do 30 years before.

‹ 27 ›

Mentors, advocates and personalities

From them you learn that they too have doubts; that they sometimes get their mathematics all wrong; and that, if of the experimental persuasion, that they have sometimes made spectacular stuff-ups in interpreting their data.

Good scientists are a product of nature and nurture. First, by nature they need to have the sort of brain that stacks facts in neat little piles and can then recognise inconsistencies or new concepts from them. They must also be stubborn and brave enough to take the 'untrodden path'—to be at odds with the prevailing truth.

However, science is a rough sport. You are always subject to severe criticism. New ideas are regarded with enormous suspicion. Even if you are right, the doubters may later claim that it was their idea and that your contribution was of minor consequence. This is where 'nurture' is important. Encouraging university lecturers are useful—but not crucial. The nurture that makes all the difference comes after that, often provided by association with kindly, well-established scientists, who are at home in the rough and tumble of the scientific world.

You learn from these scientists that they too have doubts; that they sometimes get their mathematics all wrong and, if of the experimental persuasion, that they have sometimes made spectacular stuff-ups in interpreting their data.

Over lunch, in informal 'bull-sessions' and in conversation while walking down a corridor, you gain an insight into the real world of pioneering research, and you gain that most important attribute: confidence.

In this process of informal interactions you gain access to worldwide scientific networks. Some of the people you meet become mentors, who help you through the many problems that are never in the textbooks. More important still, some become your advocates and you may never even know about it. If they look critically at your research work and future proposals, gauge them against the work of your contemporaries and like what they see, they let their networks know. This can rapidly assist with your acceptance into the international research community.

Arriving in Boston in 1959, I joined the research group at MIT that was gathered around Bruno Rossi (Plate 7). That was undoubtedly the greatest stroke of luck I could have had. It provided me with 'nurture' of the most powerful form possible.

I have spoken of Bruno Rossi before. He was one of the world leaders in cosmic ray and space research. It was an exciting, and frightening, prospect to be working in his group, even if one of my early encounters with him was not auspicious.

Two months after I arrived in America, Bruno suggested that I go to the Mid-West Cosmic Ray Conference, in Iowa City. While the name suggests a little, local meeting, this series of conferences had been central to the development of space research over the preceding decade. The previous year Jim van Allen and his co-workers had discovered the radiation belts that now bear his name. He as well as many others were going to be airing their recent pioneering research for the first time.

On returning from the conference, Bruno asked me to brief him and several other high-powered people on the latest discoveries by van Allen. I was totally out of my depth: my brain had been in a perpetual state of overload in Iowa; I hadn't understood the measurements; a lot of the words they used were new to me and I certainly did not understand the theory.

I didn't have the nerve to say any of this, of course. Still thinking like a cash-starved Australian, I couldn't admit that the cost of my airfare halfway across America had been wasted.

I tried to muddle through and started to summarise the results. I immediately made an enormous error. I said that van Allen had shown that the radiation belts were populated with electrons with energies of 40 million electron volts (abbreviated as 40 MeV). This was wrong by a factor of a thousand; I should have said 40 thousand electron volts (40 KeV).

Now physics isn't always a precise science. Sometimes you can be a factor of 1000 wrong and no one will notice it. I was not that lucky. This was a 'stuff-up' of a spectacular nature and there was a very good reason that I should have known it.

To explain, a 40 MeV electron is almost moving at the speed of light and it would quickly lose its energy, radiating out what is known as 'synchrotron radiation'. It would not stay that energetic for long at all. Radiation belts populated with particles of that energy could not exist.

A 40 KeV electron, however, moves at an almost languid speed. It does not radiate energy. It would stay in the radiation belts. Unfortunately, by not noticing the implausibility of what I was saying, I was showing a great lack of 'physical intuition'—one of the most valued attributes of a physicist

Unfortunately, by not noticing the implausibility of what I was saying, I was showing a great lack of 'physical intuition' ...

Bruno gave me a chance. 'Ken, ah, MeV or KeV?'

Muggins didn't take the hint to think hard. 'MeV, Bruno.'

Bruno didn't say anything. He quietly crossed the room and rummaged around among a pile of scientific reports. He found the one he wanted, written by van Allen, and together we checked through it. We found that it was KeV.

He speculated briefly on what the implications would have been if it had been MeV. It was a marvellous lesson to see a top class scientist thinking 'on his feet'. Moreover he did it in a way that saved me embarrassment.

I would not forget either lesson.

About a year later he met me in the corridor and asked me if I would like to teach a second-year class in optics. I demurred, 'But Bruno, I don't know much optics'.

‘That doesn’t really matter’, he said. ‘When I first taught optics ten years ago, I didn’t know any either and I didn’t until I had written my textbook on optics the following year. Don’t worry, I think you will enjoy it’.

Well what could you say? I agreed. I did enjoy it. I would stay up until 3 am before my classes, making certain I could handle the questions from some of the brightest students in America. In a poll of the students at the end of the year I was even proud to be voted one of the best instructors. Bruno himself gave me the news.

My first year at MIT was extremely productive. The Sun was aiding and abetting my research in a quite spectacular fashion, and my calculations were earning me a good reputation. This was all noticed by Bruno, who would occasionally drop into my office to ‘have a little chat’.

The ethos of the time and the large financial support available for research, meant that ‘self starters’ who had ‘good scientific judgement’ were highly valued. It is clear that Bruno was letting his network know that I had both abilities.

Soon there was a new development. The MIT group had such a good reputation that senior scientists from the world over would come to visit for three to six months at a time. Bruno would often send them to talk to me and so it was that I had the chance to work with people of great international repute and influence, such as Beppo Occhialini from Italy and Vikram Sarabhai from India.

Meeting Beppo Occhialini was a fascinating experience. Twice in his career he was the assistant to a scientist who later received the Nobel Prize for Physics—for work Beppo did with him. Younger researchers would joke, 'If you want to win the Nobel Prize, get Beppo to assist you'.

In 1960 Beppo and I became interested in using a semi-conductor—the material computer chips are made of—to detect cosmic rays. We worked out that we would be able to identify the charge and atomic number of cosmic rays that way.

Together we engaged in what Albert Einstein used to call a 'gedunken' experiment, a 'thought' experiment. We argued that the semi-conductor would allow us to measure the charge on the 'heavy' cosmic rays, and therefore identify what element they were. Helium and carbon cosmic rays had already been identified and some iron, but we wondered whether we could recognise elements right up to the top of the Periodic Table. The answer was yes.

At that time the physicists and astronomers of the world were arguing about whether the universe was created in a 'Big Bang', or whether there was 'continuous creation'.

So Beppo and I read up on the physics of the Big Bang and concluded that it would have left an indelible signature in the nature of the very heavy cosmic rays (that is, those heavier than the transition elements). We reasoned that, if we had a big enough detector and could observe from a satellite for long enough, we could tell whether the Big Bang had really happened.

Beppo would regularly drop into my laboratory to discuss all this. He would walk back and forth arguing, then, after 30 minutes or so, he would leave. I would then have to examine all flat surfaces in the laboratory carefully, particularly my laboratory bench. This was because Beppo was an inveterate chain smoker.

He was, in fact, in a league of chain smokers the like of which I have never seen before, or since—it was quite common to see him with two cigarettes alight at once, one in each hand. Worse still, he would often put down a lit cigarette, then light another, assuming he had finished the first. There was an ever-present risk of fire on my desk or of burning myself badly after our arguments. (Beppo did, in fact, live to a healthy old age and I have not yet become a victim of passive smoking.)

He would often put down a lit cigarette, then light another, assuming he had finished the first. There was an ever-present risk of fire on my desk …

Back to our gedunken experiment. I built a prototype instrument, but had to scrounge the semi-conductors at its heart, as they were not yet available for purchase. We also worked out that, while we certainly could identify the heavy elements in the cosmic radiation, it would take over a hundred years to assemble enough data to say yea or nay to the Big Bang. So we dropped the idea.
There were plenty of other interesting things to look at! However, I received a marvellous lesson in experimental design that has been of immense value to me during my career.

After his stint at MIT, Beppo returned to Italy, where he was deeply involved in the Italian space program. Italy later named one of their satellites *Bepposat* in honour of him.

Vikram Sarabhai was a remarkable colleague who I first met at MIT. Until the mid-19th century, science had been the province of talented enthusiasts, gentlemen—very rarely ladies—of independent means. They didn't call it science then; it was natural philosophy.

(I have always wondered if there was an unnatural philosophy.) Anyway, Vikram was a natural philosopher of the 20th century.

Vikram Sarabhai was the scion of a rich Indian industrialist. His father had made a considerable fortune from his Ahmedabad cotton mills, some 400 kilometres (250 miles) north of Mumbai. He had used part of his fortune to educate his son at Cambridge University and to support the Indian nationalist, Mahatma Ghandi.

Vikram hadn't worked under just anybody at Cambridge—he became a student of Ernest Rutherford, the New Zealand father of experimental nuclear physics. Vikram was very well connected from a scientific point of view.

Armed with a doctorate from Cambridge, Vikram returned to Ahmedabad, to run the family cotton mills. In his spare time, he decided to set up a laboratory to study cosmic radiation. To 'keep his hand in', he visited America each year to work for several months at MIT.

Vikram arrived for one of his annual visits not long after I arrived in 1959.

Now, I admit that our first meeting did not greatly excite me. The results he had so far published in the scientific literature were, to be kind, sloppy. I operated the same type of equipment he was using in the tropics and knew what could and would go wrong.

I was strongly of the opinion that he and his assistants did not have a 'feel' for experimental equipment and that much had gone wrong without them realising it.

This view had been strengthened by an early experience at MIT. One of the PhD graduates from Vikram's laboratory was assisting me in the operation of the abominable cosmic ray telescope I had inherited. Needless to say, I had suggested that he should do the day-to-day babysitting of the beast. Several days later he came into my room and said, 'Ken, one of the vacuum tubes is faulty'.

My response was something like, 'Oh, yes, I told you it would happen very frequently'. There were hundreds of vacuum tubes in the equipment and some bird brain had decided to use experimental tubes that had never made it beyond the experimental stage. Unfortunately, we were stuck with them.

I returned to the calculations I was doing. Several minutes later I realised he was still standing as if waiting for something. 'Is there something we need to discuss?' I asked.

'Yes, Ken, the faulty vacuum tube. Will I replace it?' I thought of all the suspect data I had seen from the laboratory in Ahmedabad, and decided I had worked out a cause behind it. If a scientist with a PhD had to be told to replace something as trivial as a faulty vacuum tube, there was a pretty big problem somewhere.

This coloured my attitude towards Vikram, but it rapidly grew worse when he started telling me all the things he wanted me to do for him—within minutes of meeting him for the first time. I'd heard he had a reputation for this sort of behaviour and, on his arrival, he demanded that the several Indians in the MIT laboratory drop their own research work to do his bidding.

I quietly explained to Vikram that was not what I was there for—and instantly our relationship changed. We became good friends. I always found his conclusions about the properties of interplanetary space to be arbitrary and unconvincing and told him so. He clearly took a lot of notice of my criticisms. Soon he invited my family and me to visit his laboratory in Ahmedabad for several months. In 1970 we did just that.

It was then that I saw Vikram's strengths and influence for the first time. The penny really dropped when Vikram introduced me to the Prime Minister of India, Mr Jawaharlal Nehru.

Vikram was not just an average cotton mill owner or senior scientist, this guy had a close relationship with the movers and shakers of a large—and growing—nation.

The penny really dropped when Vikram introduced me to the Prime Minister of India, Mr Jawaharlal Nehru.

A little research showed me why. In the 1920 and 30s, Mahatma Ghandi lived in Ahmedabad, where he started some of his spectacular acts of civil disobedience against British colonial rule. When he ran out of money, Vikram's father supplied the large sum of 12,000 rupees—which had the purchasing power of about US$150,000 in today's money.

This philanthropic funding made it possible for Ghandi to continue his campaign for independence. Prime Minister Nehru had been one of Ghandi's closest associates during those days.

Clearly Vikram's political connections were immeasurably more powerful than those enjoyed by any other scientist I have ever known—and it showed.

When I first met him at MIT, Vikram had a vision: to use television beamed from satellites to teach birth control to the inhabitants of India's 500,000 villages. It was an unusual idea to couple together space research and birth control, but soon he had the American space agency, NASA, supporting him. They lent him a fairly clapped out, but still very useful, television broadcasting satellite, the Advanced Technology Satellite (ATS6) and 'parked' it over India.

On the strength of this and his political connections, he received the money to put television receivers in 2000 villages in India—twenty years before there was television anywhere else in India, and 30 years before television was beamed from space, as is now common.

At this time even a telephone call across an Indian city was difficult, and between cities a miracle. Vikram was really 'thinking outside the box'. Television programs on birth control, hygiene, farming practices and similar subjects were made, with accompanying sound in the fourteen different languages in use throughout India. It was a pioneering example of how space technology could be used to aid the billions of people in the Third World.

Sometime later I visited the Indian Space Research Organisation (ISRO) in Bangalore, southern India. I was shown the radio reception equipment that was to be used to receive the transmissions from their rockets as they soared into space. 'Guess where we got it all from, Ken?' a close Indian friend asked me. He said they had bought it all from Woomera, for a ridiculously small sum of money. The Australian government, instead of supporting space research in Australia, had decided to sell it all off at junk prices.

In the late 1960s Vikram was appointed by the Indian government to head both their Atomic Energy Commission and the Indian Space Research Organisation. The Indian space program went from strength to strength. Thirty-five years later, India is an important space power.

When a remote sensing satellite the Americans were launching in 1995 ended up in the sea instead of in space, they arranged to obtain the photographs they needed from a satellite built and launched by the Indians. In the process they admitted that the photographs were better than those their own satellite would have given.

Vikram died in 1971, but his vision and determination lives on. Today, the Vikram Sarabhai Space Centre in Bangalore employs some 2500 engineers. They have built and launched 43 satellites, most of them communications and remote-sensing satellites of their own design and manufacture. They build their own rockets and sell communication satellites to other countries.

During a visit in 2006, they discussed with me their plans for an Indian astronaut program. True to Sarabhai's example of thinking outside the box, they discussed with me the implications of recent measurements of the radiation in space—and some of my discoveries—for intercontinental air travel and for the ongoing changes in the Earth's climate.

There's another image that pops up in my mind whenever I think of Vikram. While my family and I were living at his laboratory in India in 1970, the Chinese launched their first satellite. The Indians were horrified. *The Times of India* went ballistic, with banner headlines and talk of impending doom. India and China had been having wars in the Himalayas for several years, and they were clearly alarmed that the Chinese were well on the way to having the capability to launch an atomic bomb into space.

I clearly remember arriving at the laboratory two days later and seeing a long line of fancy cars parked where there would normally have been several ox carts and a few rather battered motor-rickshaws. There were soldiers everywhere, their 303 rifles (just like the one I had when I was in the army) very much in evidence.

As I walked to my office I saw a number of groups of Indian 'heavies' in earnest conversation. There were recognisable politicians, army and air force brass and many others.

They spent the whole day cloistered in secrecy. Six years later the results of that day became known to the world, when India exploded her first atomic bomb. Soon after that they were building their own rockets, which would allow them to throw their bombs into other people's backyards.

And so, Ahmedabad, the city made famous by Ghandi's peaceful resistance was instrumental in India's acquisition of the ultimate weapon.

The head of their Atomic Energy Commission, Vikram Sarabhai, who had been a leading participant in the so-called 'Pugwash Conferences', which aimed to restrict the use of nuclear weapons, had inadvertently initiated that process.

To me, it was a fascinating example of how truth can be stranger than fiction.

‹ 28 ›

Little green men and other weird tales

Unless something was done promptly, the Sydney newspapers would soon have headlines screaming 'Atomic Bomb Found In North Ryde'. It would disrupt our work for weeks.

'Oh, you are a space scientist. Tell me, have you seen any flying saucers out there, or little green men?'

That question is one of the occupational hazards of being a space scientist. In a sense, I have, several times—but not as the questioner would think.

The most famous occasion was in 1968, when the early risers in Sydney rang the local radio stations to report that a bright object could be seen in the sky immediately over the city. It was moving north at a rapid rate and many observers realised that this meant it was not Venus, the morning star. The more careful callers called it a UFO (unidentified flying object), as was the jargon of the time. However, some callers and early morning radio comperes didn't leave it at that. Within minutes it was being called a flying saucer.

The excitement continued to mount. Soon it was reported to be flying at a speed of 3000 kilometres (1900 miles) per hour.

The early risers in Sydney rang the local radio stations to report that a bright object could be seen in the sky immediately over the city. It was moving north at a rapid rate ...

At the time I was in Mildura, 850 kilometres (530 miles) to the west, and I was very upset. The previous day my students from the University of Adelaide and I had launched a very large balloon, carrying an astronomical telescope. It had flown slowly east at an altitude of 35 kilometres (21 miles), just as planned, and was due to be 'cut down' just before it reached the east coast. A radio signal would cause an explosive device to cut the connection between the balloon and the telescope beneath.

Usually the telescope would then fall the 35 kilometres to Earth, slowed in its descent by a large parachute. The balloon would be very cold, about minus 70 degrees Celsius (-95°F), and the shock of losing the 500-kilogram (1100 lbs) telescope would cause it to shatter to bits. The remnants would fall quickly, usually beating the telescope to the ground.

This time it hadn't happened that way. The balloon-flying contractor (an arm of the Australian government) informed me that the explosive charge had not worked. The balloon was drifting north-east over Sydney at 30 kilometres (19 miles) per hour. Soon it would be over the Tasman Sea and, when it did ultimately come down, our very expensive telescope would descend into a watery grave.

Without a doubt, the UFO over Sydney was my balloon. We knew precisely what it looked like up there. It was a large sphere, about 100 metres in diameter, made from thin polythene film. In the sky above Sydney it would have appeared bright, catching the morning sun before the light hit the city itself. It was, however, going at 30 kilometres per hour; not 3000 kilometres per hour as reported by some.

Of course, facts are never allowed to spoil a good story. The radio stations had a field day. There were updates on the position of 'the flying saucer' throughout the morning. Near noon it was reported to pass out to sea near Newcastle. It was speculated that the air force tried to shoot it down, as the American air force had tried to do, unsuccessfully, in a similar episode some years before.

Luckily for us, the telescope had, in fact, been released as hoped. It was found in a farm paddock. Sydney had its flying saucer story because the balloon had chosen not to shatter.

Not long after the balloon incident, radio astronomers in the UK made a totally unexpected discovery. They found a 'radio star', which was sending a series of pips, almost as if it were sending a message using morse code. Scientists give stars descriptive names. For example, the first X-ray star in the constellation Sagittarius is called Sagittarius XR-1. For a while the radio stars that were sending out series of pulses were given the names LGM-1, LGM-2 and so on, in which LGM stood for little green men. Unfortunately, the name did not stick. We call them 'pulsars' now.

The tirade in Spanish rolled on and on. The speaker had been going for ten minutes and was becoming more and more excited and strident by the minute.

I listened to the simultaneous translation on my earphones. The speaker was asserting that Ecuador owned about half of the 'geosynchronous orbit'—the satellite orbit that communication satellites occupy. In particular, he was claiming all the parts of the orbit occupied by American communication satellites, both civilian and military. This was an enormous claim and, if upheld, would have had huge implications for the television and communications industries, not to mention the military.

I looked around. There were signs of complete boredom. Delegates seemed to be asleep or they were chatting. The American delegation had left en masse and was gathered in a rugby scrum at the back of the auditorium. As far as I could tell, no one—not even the Americans—was taking the slightest notice of what the honourable representative of Ecuador was saying, but it didn't seem to worry him in the slightest.

After 25 minutes he sat down. The chairman called the next delegate to speak. Not once in the next nine days were the Ecuadorian claims mentioned. No one argued against them. It was as if the Honourable Delegate from Ecuador had never spoken.

This was my introduction to the surreal world of international politics. I was representing Australia at the annual meeting of the scientific subcommittee of UNCOPUOS, the United Nations Committee on the Peaceful Uses of Outer Space.

More than 50 countries were represented, and there were 100 or more delegates from around the world. Then there was the legal subcommittee, and, of course, the main committee.

Not once in the next nine days were the Ecuadorian claims mentioned. No one argued against them. It was as if the Honourable Delegate from Ecuador had never spoken.

The committee met for two weeks each year at the UN Headquarters in New York, then there was a three-week meeting in Vienna to modernise the rules for the use of space. The Cold War was still in progress and China and the Soviet Union had developed great differences of opinion.

Countries such as Upper Volta (Burkina Faso) and Ecuador were entitled to the same representation and speaking rights as the large space powers. As a consequence, the progress of the committee made the average glacier look like a racing car. I was bored stiff. Luckily, after I had served on the committee for two years, an Australian government department took exception to the fact that the country was being represented by a *space* scientist and they replaced me.

There was one memorable occasion that stands out in my mind, though. The Soviet Union had been flying large radar satellites to keep track of the movements of the US Navy. The satellites carried large radioactive sources to give them the electrical power they needed. To see the ships more clearly, the satellites were in very low orbits. The idea was that after the satellite had fulfilled its purpose, a little rocket would be fired to put the satellite and its radioactive cargo into a higher, safer orbit. Otherwise the friction of the atmosphere would cause the satellite to crash to Earth after several months.

However, it didn't quite work like that. In the first incident, radioactive material was dumped in a 3000-kilometre (1900 mile) long trail across the Canadian tundra. The Canadians were not happy. The Soviets blamed 'technical factors' and said that it was quite impossible that it would ever happen again.

With impeccable timing the Soviets decided to send another radar satellite into their cemetery in the sky two days before our committee was due to meet in New York. It seems that the technical factors had not been solved just yet. The satellite went into an even lower orbit ... and was expected to return to Earth on the day our committee convened.

John Carver of the Australian National University was the chairman of the committee. I arrived with him at the UN before the meeting where we met the senior Soviet delegate. He was treating the whole thing as a great joke.

'Next time, John, I will personally arrange that *Sputnik* re-enters over Australia and lands on your office.'

When called to speak in the meeting, he stated that, 'Superior Soviet technology had been used to cause the satellite and its radioactive cargo to return to Earth in the southern Pacific Ocean, far away from any human habitation'.

'Next time, John, I will personally arrange that Sputnik *re-enters over Australia and lands on your office.'*

He was right about where the satellite landed, but the 'superior technology' was almost certainly complete bullshit; I believe the Soviets simply got lucky.

A scientific laboratory moves to its own rhythm. It has little crises rather unlike those in other walks of life. Here's a story to give you an idea of this.

By 1974 I was working for CSIRO in Sydney. On arriving at my office one morning, my secretary told me that an urgent meeting had been called at very short notice and I was needed—*fast.*

The world was at that time experiencing its first 'Oil Shock', when the oil producing nations were holding the rest of the world to ransom. The price of oil had skyrocketed. The media played up the possibility that we were all about to run out of oil. Politicians around the world liberated large quantities of money for scientists and engineers to determine how we could make our reserves last longer. The age of recycling, and alternative energy was upon us.

CSIRO leapt aboard the bandwagon with great relish. There was research into how we could produce energy from the Sun, pig manure and wave power. There was other research into how we could save energy with more efficient technology in things like motor cars, smelters and water heaters. There was recognition that the rubbish tips of our cities were important tools in both saving and generating energy. So the rubbish tips of Sydney were measured, probed and generally annoyed.

Old-fashioned rubbish tips then had a culture of their own. They were invariably presided over by wizened old men, with a small, dilapidated shack in the middle of the tip. These were their kingdoms, and they ruled with an iron fist.

There were many rules, but the firmest was that you had to see him before you went anywhere in the nether regions of the tip. These old guardians were garrulous, so there was no such thing as a quick visit.

One of our scientists had gone to the tip the previous day. The guardian was waiting for him.

'Look here, I found this on the tip yesterday.'

He pointed to a hemisphere of grey metal.

'It's very heavy for its size,' he continued, 'At first I thought it was lead, but it gave off orange, blue and purple sparks when I scraped it with a file'.

He added that he had shown it to 'the professors from Sydney University', who were also into rubbish tip research. He said they couldn't identify it. It was clear that he had a very low opinion of scientists in general, and was quite certain that CSIRO would fail the test as well.

When I arrived at the meeting room that morning, the hemisphere of grey metal was in the centre of the table. The scientist who had found it was starting to explain the situation. It weighed five kilograms (11 lbs) and had a density of 20 (times that of water).

I sat up. 'Surely, you're wrong', I said.

He smiled. 'No, Ken. We analysed it. It's pure uranium.'

'Shit! Where's the other one?' I thought to myself.

Let me explain. I was a teenager during World War II, when the so-called 'Thin Man' atomic bomb was invented, consisting of two hemispheres of pure uranium. One hemisphere was in a cannon, which forced it to hit its companion in a hell of a hurry. Separate they were innocuous lumps of metal. Together they went off like 20 thousand tonnes of high explosive.

The other type of atomic bomb was called 'Fat Boy', which used plutonium wrapped inside a large spherical chunk of explosive.

Clearly half a 'Thin Man' atomic bomb was not meant to be in a rubbish tip in the middle of the stockbroker belt of Sydney.

If, however, it wasn't half an atomic bomb, what the hell was it and what was it doing on the tip? I knew (then) of no other use for a hemispherical chunk of uranium, roughly half the critical mass used in the atomic bomb. Given the range of possibilities, the half a bomb option almost looked the more likely.

If, however, it wasn't half an atomic bomb, what the hell was it and what was it doing on the tip?

The one thing I did know was that unless something was done promptly, the Sydney newspapers would soon have headlines screaming 'Atomic Bomb Found In North Ryde'. It would disrupt our work for weeks.

Being the senior person at the meeting, I asked everyone to speak to no one whatsoever about what we had found.

It seemed obvious that the Australian Atomic Energy Commission south of Sydney might be the people to defuse the problem, so to speak. Quick research showed that they had the legislated responsibility for nuclear material and that they could not refuse to become involved. Within minutes, we asked them to come and take it.

Some time later they filled us in on the story. There are three different versions of the uranium atom; only one will explode and it comprises only one per cent of the uranium found on earth.

The nuclear industry sometimes separates uranium into the version that goes bang (^{235}U) and what is known as 'depleted' uranium (primarily ^{238}U).

On investigation our lump was found to be 'depleted' uranium. Depleted uranium has few useful attributes, except for the fact that it is 33 per cent heavier than lead, so it was used for counterbalances, as keels for yachts, as well as in armour-piercing shells for the military and as a mass to absorb nuclear radiation. Our lump had been used for the last of these—in an X-ray machine.

How it ended up on the rubbish tip is a mystery.

‹ 29 ›

A better banknote

The bank was concerned about the ability of people to use the newly introduced copying machines to counterfeit banknotes. Several months earlier there had been a great deal of publicity and dire warnings about a batch of 'near perfect' forgeries …

The phone rang. 'Professor McCracken?'

'Yes.'

'Professor Ken McCracken?'

'Yes, that's right.'

'You are in the physics department of the University of Adelaide and have worked for NASA in America?'

I wondered, 'What's this guy up to?' It didn't improve either.

'Could I come and discuss something with you?' he said.

'What will it be about?'

'Oh, I'm sorry, I cannot tell you that.'

'Well, who do you represent?'

'I can't tell you that either.'

The first James Bond movies were very much in vogue. I had the distinct feeling that one of my weirder friends—and I had a fair few—was trying to pull my leg. Nevertheless, he eventually convinced me to agree to meet with him. To be honest, I did not expect this person to turn up.

But turn up he did, middle aged, distinguished and looking very right and proper. Before saying where he was from, he stressed that the meeting must be in total confidence and that I must not tell others where he was from or what we would talk about.

After agreeing to that, he introduced himself as being from the Reserve Bank of Australia. He told me that the bank was concerned about the ability of people to use the newly introduced copying machines to counterfeit banknotes. Several months earlier there had been a great deal of publicity and dire warnings about a batch of 'near perfect' forgeries, which were released onto the market in the last few frantic days of selling prior to Christmas.

What was not said, but I soon learned, was that they were really excellent forgeries, yet they had been made using a very cheap and widely available copying machine. They were only detected because the notes didn't feel quite right to the touch. The Reserve Bank, which prints our money, was going off its brain. Next time, they thought, someone will achieve the right feel.

They had good reason to worry. Until then forgeries required great skill and much time to make. The forger had to make a metal printing plate, which would be used in a printing press to make banknotes one by one. Every single detail of the banknote had to be engraved in the metal perfectly. To make the job harder, banknotes were covered by an intricate pattern of swirling lines and patterns. Only the very best engraver would be able to do the job—and it would take months. The bank reasoned that the intricate patterns was an almost impenetrable defence against forgery.

As a second line of defence, each banknote had a different serial number. The forger, even if he had an excellent engraving, would not have the specialised equipment to change the numbers on each forgery. As a consequence, all the counterfeit notes would have the same serial number. As soon as a forgery was detected, the bank would advertise the number on the counterfeits and they became dangerous to use. The limitations of the forger's equipment meant that the 'use-by' date of a forgery was only a few days.

Imagine the consternation in the Reserve Bank with the introduction of the photocopier. All the wiggles and other secret protective devices built into a banknote could now be reproduced with a simple piece of office equipment and hundreds, even thousands, of different serial numbers could be used—too many for a shopkeeper to check. The consequences were too horrible for the bank to contemplate.

All the wiggles and other secret protective devices built into a banknote could now be reproduced with a simple piece of office equipment …

Herbert Cole 'Nugget' Coombs was the head of the Reserve Bank of Australia. He was a man of great intelligence, understanding and ability. He saw that protecting banknotes was now a role that should be delegated to the capable hands of science.

I was subsequently asked to sit on a panel of technical people whose task it was to think of a way to keep ahead of the forgers.

First we visited the Mint in Melbourne. We learned about engravings, about the printing presses, the inspection process, etc. We learned about the 'lifetime' of a banknote and how much each banknote cost to produce.

Then we visited their 'rogues gallery', where they kept examples of the work of forgers in the past. We saw the products of 'Mister One by One', who had painstakingly used a pen and red ink to manufacture 10 shilling notes; then the examples of forgeries using an engraved plate. We saw that the engraving always had quite obvious errors, once you knew where to look. We soon understood the seriousness of the threat posed by the office copier.

Our committee met several times, first at the Mint in Melbourne and later at the Perisher Valley Resort in the ski fields of New South Wales.

I knew I was operating in a new league when I checked into my room. Along with the instant tea and coffee, there was a 700-millilitre bottle of Johnny Walker whisky. Checking with my colleagues, I found that we had all been supplied with this form of refreshment, to help us think in creative ways, I presume.

We discussed many technical tricks that might be used to foil the counterfeiters. It had to be cheap to manufacture, easy to use and the banknotes would have to have a long lifetime. We discussed using special papers and plastic. We discussed electrical tricks. We spent a great deal of time discussing optical tricks; the use of interference fringes that would change colour as they were moved was discussed at great length, as were moiré patterns and even holograms.

Our deliberations finished and our report written, the Reserve Bank contracted a Melbourne Division of CSIRO to develop and test the ideas we had put forward. Almost twenty years later the first of the new Australian banknotes were issued. Many of the ideas discussed in the NSW ski fields were clearly evident.

Our committee and the combined efforts of the Reserve Bank and CSIRO initiated a revolution in the production of banknotes. The majority of countries have followed Australia's lead and use plastic notes with features that a copying machine cannot reproduce. A number of countries have their currency printed in Australia using our techniques.

My lack of 'scientific baggage' has often been seen by others to be a major advantage. For example, I was involved in some space-related work in the Philippines in 1990 when I received a phone call from the Melbourne head office of the mining company BHP.

They asked whether I would assist them discover whether the Earth's gravity could be measured very accurately from a small aeroplane flying at an altitude of 100 metres (330 feet). To some this appeared to be totally impossible; some said against the laws of physics. I explained that I knew nothing about measuring gravity.

They replied, 'Yes, we know that, which is exactly why we want you to assist us. We want to take a fresh look at it without any of the baggage of the past'.

That project is beyond the scope of this book, but it achieved what was widely thought to be impossible. I regard it as one of my most challenging, difficult and satisfying projects. I am convinced that I would never have been asked to participate if it were not for my background in space research and my acquired ability to take big risks.

Due to my space research I had developed a reputation of being able to solve the myriad problems—many small, a few big—that inevitably beset pioneering science and technology.

The pioneering days of space may be over, but those acquired skills are now being used in many other important ways in Australian industry and society.

‹ 30 ›

The big change

Soon after we reached Tasmania, I received a phone call from the chairman of CSIRO. He asked me if I would meet him in Melbourne to discuss some ideas about the application of physics to the Australian mining industry. I made it clear that I would be leaving Australia soon, to which he replied, 'Yes, we know that'.

By 1970 I had been working in space research for eleven years. I was, however, finding it extremely hard going to continue that sort of work in Australia. There was little assistance for space science in the universities and nearly all of my active involvement was through my association with the space programs of other countries. I was leading several investigations in America and our rocket flights from Woomera were conducted as part of the British space program, which was about to terminate its activities in Australia.

This was not a satisfactory situation and it clearly did not have a great future. Some serious thought led me to two options: if I were to remain in space research, I would have to leave Australia; or if I wanted to stay in Australia, I would need to change into another field of science.

Unlike many of my contemporaries in science, who might spend their whole scientific career doing the work they did for their doctorate, the idea of moving into another field caused me no concern whatsoever. In fact, I had been thinking that a change would be a good thing for several years, in an abstract sort of way. I had already changed from my initial choice of career—industrial chemistry—to a form of nuclear physics and from that into a form of astronomy.

Anyhow, scientists are great gossips. Soon my quandary came to the attention of a number of the leaders in space research in America. Owing to my 'can-do' attitude and proven track record in extracting big bucks from NASA and other agencies, I soon received attractive offers from America.

In particular, I received an offer of a full professorship at MIT. Having spent three happy and productive years there, ten years previously, I was tempted. We had many good friends there; it was one of the best technical universities in the world and it was very big in space research.

I pondered. I could see no good opportunities in other fields in the universities of Australia and enquiries into opportunities in radio astronomy and conventional astronomy were fruitless, so the decision was made. We were going back to Boston. We sold our Adelaide home and went to spend a few weeks with our families in Tasmania before leaving.

Soon after we reached Tasmania, I received a phone call from the chairman of CSIRO. He asked me if I would meet him in Melbourne to discuss some ideas about the application of physics to the Australian mining industry. I made it clear that I would be leaving Australia soon, to which he replied, 'Yes, we know that'.

At that time, I knew little about CSIRO. It had a wonderful reputation and I didn't regard myself as good as many of the people in it. I set off to Melbourne, somewhat bemused, but determined to have an interesting day.

I took a taxi to the headquarters of CSIRO. (This was before they moved to their grand address in Canberra, often referred to in the organisation as Tombstone Territory). The taxi delivered me to an unimpressive building, up a scruffy-looking alley. I thought, 'This is the headquarters of CSIRO, one of the most respected scientific organisations in the world?!'

It certainly was. Inside I received my next shock: I was met by the chairman, who introduced me to the other members of the executive board. Clearly, this was not just a courtesy meeting.

We immediately got down to business. They talked about the previous year when Australia had been gripped by a frantic nickel boom. Apparently their contacts in the mining industry were extremely unhappy with the quality of the exploration instruments being sold by the enthusiastic and plausible salesmen from large and influential overseas corporations. They promised much—and the stock market loved the idea that science was finding the minerals no one else could find—but the results did not match the sales pitch.

They asked me for my ideas on how space science techniques and ideas could be used to make the exploration for minerals more exact. I was ready for this and explained how my space colleagues and I had been participants in the early 'digital revolution'. The use of the predecessors of the computer had allowed us to digitise our measurements and given us enormous improvements in the accuracy and the reproducibility of our results.

I pointed out how measurements taken in the harsh environments of the Australian outback could benefit from the robust technologies that had proved to be so important in the equally harsh environment of outer space. The CSIRO executive board loved it.

We talked of the differences between Australia and the rest of the world. I had looked at this a bit. Certain geophysical testing techniques—that had caused a lot of grief during the Australian nickel boom—had been developed in Canada, a landscape which had been swept clean by a glacial ice-sheet within the past 12,000 years. When you sent electrical signals into Canadian ground, there was nothing to confuse or divert them from locating mineral deposits.

In Australia, which had not been glaciated for a hundred million of years, there were up to 100-metre (330 feet) thick layers of salty earth, which played merry hell with the electrical signals.

CSIRO asked what could be done about this. I pointed out that the problem was that the electrical frequencies which worked so well in Canada were hopeless in Australia, but that there was a way around the difficulty.

The solution was really quite simple: use lower frequencies, which would result in—using the language of the physicist or engineer—a skin depth greater than 100 metres. The physicist on the Board of CSIRO, Victor Burgmann, loved it.

They asked me what I knew about the practical aspects of mineral exploration. My succinct answer was, 'nothing'.

Victor Burgmann said, 'Great! Exactly the right answer'. They asked me to estimate how many people would be needed to achieve something worthwhile. I made a wild guess.

So now they reached the point, 'We know you are off to a senior position at MIT, but would you consider an alternative—if we asked you to set up an entirely new laboratory within CSIRO to develop physical techniques for the detection of buried minerals in Australia?'

'Before you give an answer', they said, 'we must stress that we are not certain we could give you the people or funds you would need. If you said that you were interested—we do not ask for a commitment—if you were interested, we would undertake to decide whether we could find the support you would need, and give you an answer within three months'.

Now I had a big problem. I had a serious offer from one of the top universities in the world, which would allow me to stay in space research. Now I also had this half-arsed suggestion of a possibility of an offer that would allow us to stay in Australia. The latter did have the advantages of being associated with the CSIRO and one of the major industries of Australia, which, in my mind, was of equal significance as the American space industry.

Was it going to be a real offer? More to the point, would I have the courage to stop doing the things I was experienced at and plunge out into something I knew nothing about? So I mumbled something about it being interesting and they stressed that they were not making a firm offer.

My family and I then went off to spend five months working in the Indian space research program. I read books on mineral exploration. I wrote scientific papers and a bestselling book on metric conversion. However, there was the ever-present question of which option I should choose.

After about four months I heard a great commotion outside my office. One of the Indian secretaries came in and gasped out that there was a long distance call coming through for me from Australia in ten minutes. We had only one telephone for long distance calls, so I went downstairs to the administrative office to wait for the call.

The Indian telephone system in 1970 was a little less than perfect. A phone call across the city was an adventure. A call to Bombay (Mumbai), 400 kilometres (250 miles) away, was almost impossible. Under the most favourable conditions, you would hear a faint little voice, barely audible through a cacophony of clicks, whistles and a noise like waves crashing on the beach. The voice was usually saying, 'Hello, hello?' endlessly, because the owner could not hear anything. Generally, however, there was no little voice.

So I went to wait for my phone call. I waited for two hours. Finally the telephone operator called to say that the line was too busy and that the phone call would occur at the same time the following day.

The next day was the same. As before the call was postponed another day then another. For four days the chairman of CSIRO tried to call me. Each time he could not get through.

Finally he sent a telegram offering me the job of setting up an entirely new research activity in CSIRO. When we met some months later, his comments about the Indian telephone system were less than charitable.

The decision was made. We would return to Australia.

Then the great 'taxation clearance' episode started.

In those days it was always necessary to obtain a taxation clearance before travelling overseas from Australia, to make certain you did not escape without paying your tax. Therefore being told that I would need one before leaving India was not a shock.

What was a shock was that my wife, who had not been employed in India, would need one also. However, life in India was always a little surprising, so this was normal enough in the scheme of things.

I was told that it would take a week to receive the clearance. That seemed excessive, but then, this was India, I thought.

That was not the end of the matter, though. My Indian colleagues explained that I would need to visit the taxation office each day for a week to 'progress the matter'. Although I thought I was well attuned to Indian ways by then, I protested that that was excessive. My colleagues simply said, 'Do it'.

I met my taxation official, Mr Gupta Sarma. I handed him the details I thought he needed. He hardly looked at them.

'So you are at the Physical Research Institute … are you a physicist?' he asked.

Once he knew I was, the rest of the half-hour interview had nothing to do with taxation. 'I worked at the Tata Institute for my masters degree', he said. (The Tata Institute of Fundamental Research in Bombay is a fantastic institute and I knew it well).

This bemused me, though, because I thought the Tata did nothing other than physics and I could not see the connection with taxation. I asked what type of research he had done.

'Oh yes, my masters degree was in the study of general relativity. I studied the curvature of space. It was very interesting.'

General relativity was the sort of work Albert Einstein did when he grew up. He revolutionised physics in 1905. Ten years later he grew serious and tried to explain everything—electricity, magnetism, gravity, atomic physics—all with the one mathematical equation. One of the simpler concepts of general relativity is that space is curved, so that if you started out from Earth in a straight line and travelled far enough, you would return to Earth without ever changing direction.

'Oh, yes. Unfortunately, we have here a matter of fundamental incompetence.'

I asked a dumb question, 'So why are you a taxation official? I didn't think the taxation department was into general relativity?'

'Oh yes, you see when the taxation department advertises for people they take the people with the highest qualification. No one else had a masters degree in general relativity, so I got the job.'

For five days we discussed little else except general relativity. This was quite a strain on me—I knew stuff all about it—but he seemed to be happy.

On the fifth day, we spoke about general relativity for 25 minutes, before I reminded him that I had tickets to leave India the following day and would very much like to have my taxation clearances so we could go.

'Oh, yes. Unfortunately, (which is a common word in the Indian vocabulary) we have here a matter of fundamental incompetence.'

'A what?'

'Oh, dear! Fundamental incompetence. Unfortunately, I am fundamentally incompetent to deal with your Mrs.'

I should perhaps explain that Indian English at this time had its own peculiarities. In even the best circles the word 'wife' was substituted with 'Mrs'. Nobody ever talked about 'You and your wife', it was always 'You and your Mrs'.

'Oh, dear! I have here your taxation clearance ...' He produced a piece of paper that looked like poor quality toilet paper, with Hindi written all over it, 'Unfortunately, (that word again) I am fundamentally incompetent to give you one for your Mrs'.

I was not impressed. I asked the obvious.

'Oh, dear! It is really quite simple. I can give you your taxation clearance because you have earned money. Since your Mrs has not earned any money, I am fundamentally incompetent to issue you a taxation clearance for her.'

I found this rather hard to understand. I pointed out that since my wife ... er Mrs had not earned any money, I believed that she would not be required to pay tax. Having complied with the laws of the land in this regard and not having paid any tax, she should be able to be given a taxation clearance. I think I suggested that someone who knew that space was curved would not have any real difficulty with this concept.

Mr Gupta Sarma clearly would have preferred to discuss general relativity. He was unhappy with the practical turn the conversation was taking. 'Oh, dear! You see that is not my speciality. You will need to go to someone who specialises in the taxation of people who do not earn any money.'

I was not happy. I asked him why he had not pointed this out to me when I first met him five days previously. I asked where I would find this specialist who handled the taxation of people who did not have any liability to pay tax. I think I asked him whether you needed a PhD in general relativity to allow you to deal with the taxation of people who did not earn anything.

Mr Gupta Sarma explained that it was currently a religious holiday for the specialist and that he was not there. Before I did my block, he hastened to say that the specialist would be at work the following day, Saturday.

I protested, 'But our plane leaves at 11 am'.

'Do not worry. I will see that you get it.'

True to his word, the second piece of toilet paper with Hindi on it arrived shortly before we left for the airport. Furthermore, it was delivered personally by Mr Gupta Sarma, riding a bicycle. Clearly I had made his week—it seems that there were few customers of the taxation office in that small city of India who could discourse on the curvature of space.

For my part, I learned a new concept—the concept of fundamental incompetence. It is a handy thing to know. Knowing about it has helped me greatly later in life when dealing with all sorts of administrators.

The only difference is that in India, being fundamentally incompetent meant that you were not allowed to do something.

In some Western countries it seems to me that being fundamentally incompetent is a prerequisite for certain positions. I think that is what is referred to as a cultural difference.

... in India, being fundamentally incompetent meant that you were not allowed to do something. In some Western countries it seems to me that being fundamentally incompetent is a prerequisite for certain positions.

With our taxation clearances in hand we were driven at reckless speed to the Ahmedabad airport. In the Indian way, the driver was continuously honking his horn while deftly avoiding the sundry collection of cows, goats, hawkers and the populace in general who preferred to walk down the middle of the road.

We flew back to Australia and, in early July 1970, I started working for CSIRO. For fifteen years I immersed myself in setting up and managing a research laboratory, which by the time I left, employed about 200 people.

‹ 31 ›

Remote sensing

Our analysis of the concept had convinced us that satellite photography would have a number of advantages over the photographs taken from aeroplanes, a common practice both then and now. Not everyone agreed with us.

Fifty and more years ago, Australian scientists were quite isolated from the cut and thrust of scientific discovery. Without the internet and reliable international telephones, and in the face of expensive and slow international travel, they had little access to the conferences or 'gossip networks' in Europe, the United Kingdom or America.

Australian science—or at least CSIRO—had an answer to this. They stationed a 'scientific councillor' in the Australian embassies of London and Washington. They were usually well-known and competent scientists or scientific administrators, who were able to quickly 'plug into' the scientific world in whichever neighbourhood they found themselves in.

While employed by the CSIRO, they saw their task as extending to the wider research community. They became the eyes and ears of Australian science overseas.

This arrangement served me well while I was still at the University of Adelaide. Before leaving America I knew that I would be receiving five or more large computer tapes each week from the satellites I had in orbit. Sending them back to Australia safely by air was going to be very expensive indeed.

Fortunately, I had come to know the scientific councillor in Washington, Clive Garrow. On one of my visits to Washington, I discussed my problem with Clive. Some days later he called me to say 'get NASA to send the tapes to me at the Australian Embassy'.

Each week a diplomatic bag was sent from our overseas embassies to Canberra, carrying all the diplomatic, political, commercial and other information that the Australian government needed. Into that bag went my many computer tapes, and having reached Canberra, they were sent to me in Adelaide by parcel post.

In addition Clive occasionally sent me letters telling me about new developments that he thought might interest me. In 1969 he was approached by an American company, asking whether Australia would be interested in 'remote sensing', that is the use of satellites to take 'photographs' of the Earth, which could be used for mineral exploration, hydrology, monitoring crops and forests, and so on.

Alerted by his letter, I did a little research. From the beginning of the Space Age, the American military had been interested in using satellites to peer into the rocket launching, atomic research laboratories and military establishments in the Soviet Union. America had been designing 'spy' satellites even before the Soviet Union launched *Sputnik 1*. They became even more interested after the U-2 spy plane was shot down over the Soviet Union in 1959. Within several years they had spy satellites in operation, which steadily improved as the years went by.

Other people wanted to get in on the satellite act. In particular, the US Geological Survey announced a plan to fly a satellite, *Earthsat*, which would take photographs of the Earth to be used for mineral exploration. They called this 'remote sensing'.

NASA took a dim view of this threat of competition—and of money going to other organisations than NASA. Much political infighting took place, but NASA was eventually victorious.

To win, they had to agree to fly a satellite that would look down on the Earth, not outwards as they had usually done before. They called it *Landsat* and it was designed to take photographs of the Earth that would be useful for many activities other than simply mineral exploration.

In 1970, several days after returning from India and joining CSIRO I attended a meeting in Canberra to discuss the useful application of *Landsat* imagery to the Australian economy. Along with four scientists from the Australian Geological Survey, National Mapping and CSIRO, I prepared a detailed proposal for Australian scientists to use the images taken by *Landsat 1*.

There was a catch, however. There was no satellite receiving station for *Landsat 1* in Australia, so images of the country were first recorded on a tape recorder on the satellite before being sent to Earth by radio when the satellite passed over Alaska. My laboratory in Sydney bought as many of the computer tapes containing this data as we could, to allow our scientists and the mining industry to experiment with their use.

Our analysis of the concept had convinced us that satellite photography would have a number of advantages over the photographs taken from aeroplanes, a common practice both then and now. Not everyone agreed with us.

Many influential members of the mining industry found my employment at CSIRO quite incomprehensible at first. 'What', they asked, 'would a space scientist know about mining?' They assumed that it was another example of the Australian government wasting money that could be better used on the ground.

Our concept of using satellite photographs—taken from a height of 800 kilometres (500 miles)—simply confirmed their idea that we were a bunch of 'academic wankers'. Their words, used to our face. They called it 'remote nonsensing'.

Our concept of using satellite photographs ... simply confirmed that we were a bunch of 'academic wankers'. Their words, used to our face. They called it 'remote nonsensing'.

Undeterred, Andy Green and Jonathon Huntington, two young colleagues at CSIRO, developed ways to greatly improve the quality of the *Landsat* images. We proposed an ambitious research program to the Australian mining industry, but at this point I made an incredibly stupid decision.

Our best *Landsat* image was of an area in the farming areas of Victoria. It showed the details of features such as roads, fences and creeks very well. I thought this would allow the geologists to see how useful the imagery could be and decided that the Victorian image would be included in the proposal.

The mining industry responded with a great yawn. Not one mining company showed any interest.

Some time later we processed an image of part of Western Australia, near Marble Bar and the Hamersley Range. It's a spectacular image that made geologists gasp, go cross-eyed and utter strange noises.

It showed a number of different types of rock in close proximity to one another, allowing geologists to predict where important mineral deposits might occur. Quite subtle differences in the geology could be clearly seen, which was barely possible otherwise.

We proposed our satellite-assisted, remote sensing research program to the mining industry again, using the Marble Bar image. There was a stampede—ten major companies wanted a piece of the action.

I learned how interested the miners had become in this form of space science a year or so later, when I went over to the *Landsat* research area to discuss something.

'Where's Andy?' I asked and was directed to the small dark room we used to examine the images on a large computer screen.

As I walked in, the light from the computer screen showed that the room was absolutely packed by people who clearly didn't want me to see them there. I recognised one of them; he was from one of our largest mining companies. Sensing that I wasn't wanted, I retreated.

I later found that the company had made a really big discovery, which was still very hush hush. All the top people in the company were in our little dark room looking at what else could be seen from space and wondering how many more discoveries they could make. They and seven other major mining companies, became avid converts to our satellite images.

Over the next few years, more mining companies became interested in satellite imagery, but in 1981 another problem developed. America launched a new *Landsat*, which could provide much better images.

Unfortunately, the receiving station that the Australian government had finally set up for the earlier *Landsat* satellites could not receive transmissions from the new satellite. The Australian Government, being short of money, said they would not pay for the modifications to make this possible.

I hatched a plan to design and build the electronic equipment required to modify the existing *Landsat* receiving station for a modest sum, and presented it to the mining industry and CSIRO. Between them, they provided the $500,000 that allowed my engineers and technicians to make the modifications.

Government sensibility prevented us using *Landsat* in the name, so we called it the Signal Processing Experiment. Funny names aside, it did the trick.

The mining industry and other researchers obtained the much-improved imagery years earlier than they would have done otherwise. It was a good example of space science being applied for practical purposes.

‹ 32 ›

Winding the clock back

We could see ... how a very large burst of cosmic radiation from a solar flare could produce nitric acid in the atmosphere, which would then precipitate to earth in snow. We could see that there were more spikes in the nitric acid, all the way back to periods before Galileo saw the first sunspots in the early 17th century. Were we looking at a record of the Sun's production of cosmic radiation for the past 400 years, we wondered?

In Chapter 7 I described solar flares and the pioneering observations of Carrington and Hodgson. Over the decades their observations were vindicated and a great deal of knowledge was gained about the nature of solar flares and their impact on Earth. Among other effects, it was learned that they would, on occasion, generate intense bursts of cosmic radiation, which could have negative effects on satellites and humans as well.

One of my first scientific investigations in 1956 was into the six known bursts of cosmic radiation generated by solar flares over the previous 20 years (e.g. Figure 1). In the same period there had been only eight solar flares seen in white light, as was the case for Carrington and Hodgson.

I saw that most of them had coincided with the known cosmic ray bursts and speculated that there had probably been a very large cosmic ray burst at the time Carrington and Hodgson had seen the strange bright line of white light. Of course, there was no way I could prove this …

That is, not until 1995 when two visiting colleagues, Peggy Shea and Don Smart, showed me the analysis of ice core samples taken in Greenland.

Let me explain. An ice core is obtained by using an instrument—similar in principle to an apple corer—to remove a very long plug of ice from a glacier or ice sheet. The top of core will have fallen as snow the previous year, but further down the ice may be hundreds of thousands of years old. By analysing the ice layers you can build a historical record of changes in the atmosphere and climate.

Regarding the case of the Greenland core, the proximity to Iceland and the existence of meticulous records of the days on which local volcanoes had erupted meant that the date each part of the ice core had been precipitated as snow was well known.

My friends pointed out that there was an enormous, one-month long spike in the concentration of nitric acid in the ice core at the time Carrington and Hodgson had seen the solar flare in 1859.

We could see, in general terms, how a very large burst of cosmic radiation from a solar flare could produce nitric acid in the atmosphere, which would then precipitate to Earth if it snowed.

Furthermore, we could see that there were more spikes in the nitric acid, all the way back to periods before Galileo saw the first sunspots in the early 17th century. Were we looking at a record of the Sun's production of cosmic radiation for the past 400 years, we wondered?

There was a big problem, however. The scientific community that studies ice cores regarded the data I had been shown with total disdain. The data had been obtained by to very innovative and careful scientists, Gisela Dreshhoff and Ed Zeller, who had analysed ice cores from the Arctic and Antarctic with much greater resolution than ever done before. While no one in the ice core community had ever made the type of measurement that we were looking at, it was stated confidently by the experts that the results *must* be wrong.

No amount of explanation would change their view. I pointed out, using elementary mathematics, that the short-lived spikes in the record would never have been seen in the coarse measurements other scientists had made. The answer given by a leader in the field was that he 'did not understand mathematics', and that, as far as he was concerned, was that. As I've mentioned, scientists can be a picky bunch when it comes to new ideas.

The answer given by a leader in the field was that he 'did not understand mathematics', and that, as far as he was concerned, was that.

His attitude stirred me up a bit. Clearly we had to prove that the spikes of nitrate were really there and that they were associated with solar flares. To do this we needed to use mathematics. Furthermore, it was clear that space researchers would be interested in the results, even if the glaciologists weren't—and space scientists are usually very good with mathematics.

I decided to conduct a comprehensive study of the results and. if the case were absolutely watertight, it would be submitted to a scientific journal that services space researchers.

It took five years part-time to complete that study, while working for the mining industry and running the farm that we had bought in the southern highlands of New South Wales in 1989.

The results were clear and we published two large scientific papers in the *Journal of Geophysical Research* proving what the spikes were and what they told us about goings on in space for the past 400 years.

It was a slam-dunk. There was no argument. Five years later our results were being used as the definitive record of solar cosmic ray bursts by those studying the hazards mankind will face when they travel to Mars.

While looking at the results from the Greenland ice core I learned about other measurements that were made from them. In particular, I became interested in a radioactive form of the element beryllium.

It was well known that an atom of this rare form of beryllium was sometimes made in the high atmosphere when a cosmic ray from space smashed an oxygen or nitrogen atom to smithereens. So, if you measured the amount of this rare form of 'cosmogenic' beryllium in an ice core deposited as snow in the year 1800, you would have a measurement of the intensity of the cosmic radiation falling to earth then.

> *This caused most cosmic ray scientists to assume that 'there was something queer about the beryllium measurements' …*

No one argued about that. However, the amount of beryllium was seen to change quite dramatically from decade to decade and century to century. The intensity of the cosmic radiation at Earth was well known from 1951 onwards and it hadn't changed anywhere near as much as the beryllium had changed up to 1950.

This caused most cosmic ray scientists to assume that 'there was something queer about the beryllium measurements', which apparently gave them an implied licence to ignore them. That the results might suggest that cosmic ray intensity in the solar system had decreased greatly since around 1900 was only considered by a few lone voices.

Someone obviously needed to prove that the beryllium results were really telling us how the radiation in the solar system changed over the centuries, and millennia, before there were satellites or ground-based instruments to measure it.

My approach was that we would think of all the things that might affect the amount of beryllium in the ice cores, and systematically determine whether they could do so to any reasonable degree. My 1960 mathematical model of interlocking calculations was one of the first used in space physics; now mathematical models are de rigueur if you want to be taken seriously in this field. I needed to develop a model for the beryllium problem.

I set about developing a model for this problem, including all the factors that might affect the beryllium levels. For example, the Earth's magnetic field varies greatly from century to century, both in its strength and in the position of the magnetic poles on the Earth. We know from other measurements how big these changes have been so we use them to 'crank in' those effects and see what it does to the beryllium.

Luckily, I was not alone in this endeavour. Two scientists in Switzerland had done the first part of the problem using complicated nuclear physics mathematical models, which calculated how much beryllium was made when a cosmic ray smashed into an atom in the upper atmosphere.

I did not have the expertise for this part. They, on the other hand, did not have a background in cosmic ray physics. I therefore took their results, wrote a mathematical model including all the cosmic ray bits and pieces and worked out how changes in the magnetic field, atmospheric circulation and other factors would influence the beryllium in the ice cores. The mathematical model showed that more than 90 per cent of the observed changes in beryllium were due to changes in the intensity of the cosmic radiation at Earth. All the other 'but what abouts' were eliminated.

Again there was no argument. Beryllium was immediately accepted by the space research community as a fair dinkum measurement of the cosmic radiation in the past. A group of researchers in Finland joined in with further proof. Some tidying up remains to be done and more ice core measurements will add to the accuracy.

With that out of the way, my European colleagues and I examined the beryllium records for the past 1100 years. To pacify the remaining doubters we analysed results from both the Arctic and Antarctic. The samples confirmed that the cosmic radiation varied greatly over that time. Between 1420 and 1540 and close to 1700 the radiation intensity in the solar system was much greater than now. Coincidentally, the Sun had virtually no sunspots during those periods.

This result caused us to look at other periods when there weren't many sunspots—near 1800 and again near 1900. Once again the cosmic radiation was greater than now.

The next obvious question was: what happens when the Sun is very spotty? The records kept by Chinese astronomers tell us that the Sun had a bad case of the measles between 1200 and 1375, then again in Galileo's time, in the mid-1700s, the mid-1800s and since 1940. Without fail, the beryllium responded each time indicating that these were periods of low cosmic ray intensity on Earth.

Using the nitrate results to tell us about the occurrence of large bursts of cosmic rays from the Sun, together with the beryllium data, we have come to the conclusion that the Sun has played a trick on us space scientists.

Our Space Age, since 1957, has coincided with an all-time high in 'solar activity'. Several complex groups of spots are often visible throughout the eleven-year sunspot cycle. For much of the previous 1000 years, the Sun has been much less active, to the point that Edmund Halley, of Halley's comet fame, wrote in the 1680s that 'he hoped that he would see a sunspot before he died'. So sunspot activity during the Space Age is the anomaly, not the norm.

There is recurrent talk about mankind going back to the Moon. The Chinese and Indians are separately considering their own manned programs. America is also conducting studies in preparation for a manned mission to Mars. Our research suggests that the conditions may be quite different from those present when mankind went to the Moon in the 1970s. The radiation conditions could be a lot worse. The historical records cannot give us a precise prediction of when this will happen, but they tell us how bad it might be. We ignore that evidence at our peril.

‹ 33 ›

To the edge of the Solar System

Finally, 26 years after it was launched, at a distance of 92 astronomical units (92 times the distance from Sun to Earth), Voyager 1 *sent back some very unusual results. There were many more energetic particles than before. The magnetic field had increased, but not by a factor of four. Controversy raged.*

We all know that the outermost planet of the solar system is Pluto (now reclassified as a planetoid), 39 times further from the Sun than the Earth, but does the solar system end there?

Using the measurements of the solar wind blowing past Earth and estimates of the strength of the magnetic fields in interstellar space, scientists estimated that the solar wind comes to a screeching halt about 90 times the distance between the Sun and the Earth (abbreviated as 90 astronomical units or AU).

However, was our understanding of the behaviour of the solar wind correct?

Was our estimate of the interstellar magnetic field correct?

An error in either would make a big difference.

Back in the 1960s, our NASA committee (Chapter 25) considered the possibility of sending a satellite to the outer reaches of the solar system. The rockets in those days, and even now, do not have enough energy to accelerate a satellite to the velocity needed to leave the solar system. To achieve this we would need to play a game of celestial billiards. By approaching the Moon or a planet from the right direction, at the right time, a satellite can use a slingshot effect to 'pinch' energy from the Moon, which makes it go faster and further out from the Sun.

On rare occasions all the planets are lined up, so that a satellite can go past up to five, pinching energy with each flyby. We had a name for that type of space mission, the Grand Tour. If done properly a satellite could go fast enough to escape the gravitational pull of the Sun and ultimately soar out into 'interstellar space'.

It took a decade or so to develop the necessary technology to embark on a Grand Tour. There were serious problems to solve. While satellites near Earth obtain their electrical power from solar cells, the strength of the sunlight falls rapidly as you fly away from the Sun. At Jupiter, it is only four per cent of that at Earth; at Pluto, the strength of the sunlight would be 0.06 per cent of that at Earth; at the edge of the solar system it would be 0.01 per cent.

Solar cells would never do. The only solution was to use a nuclear power source. In this, a large lump of radioactive metal grows hot as a result of the disintegration of the radioactive elements. The heat generates electricity using a thermocouple.

Transmitting the observations back to Earth from the outer reaches of the solar system was a major problem, too, since radio signals become weaker over distance by the same percentages for the strength of sunlight at the satellite. The only solution was to use directional radio antennae on the satellite to send all their radio waves in the direction of Earth. By contrast Earth-orbiting satellites send their radio transmissions in all directions at once.

Of course the Grand Tour satellite would then always need to know exactly where Earth was. Giant radio antennae—up to 72 metres (240 feet) in diameter—would be required on Earth to catch enough radio energy to allow the signals to be understood. These were all major technical and engineering challenges, which took many years to overcome.

Having planned a Grand Tour, NASA ran out of money and cancelled it. Jim van Allen organised a group of senior scientists who argued that several satellites then nearing completion could be launched into something like a Grand Tour. By the late 1970s four satellites, *Pioneers 10* and *11*, and *Voyagers 1* and *2*, were ready to explore the solar system out to Uranus. If we were lucky, they would go on to Neptune and Pluto.

The four satellites had television cameras on board so we could see the outer planets in much better detail than was possible from Earth, even with the largest astronomical telescopes. They would have magnetometers, to see if the planets had magnetic fields like Earth's. If they did, that would indicate they had liquid cores. All of this would tell us more about the birth of the solar system more than four billion years ago. They also carried all the usual suspects: solar wind, cosmic ray and energetic particle experiments to detect any van Allen-type belts around the planets they passed.

All four satellites were launched successfully between 1972 and 1977. Over the following years they flew by Jupiter, Saturn, Neptune and Pluto. We saw the swirling clouds of ammonia gas on Jupiter in marvellous detail. We learned much more about the

Jovian moons and were fascinated to see that the moon Io had more than eight volcanoes spewing out plumes of sulphur dioxide.

Voyaging on, the satellites sent back images of the rings of Saturn, showing in exquisite detail the hundreds of thin rings, with bead-like clumps strung around them.

As the satellites retreated from the Sun, the strength of the solar wind and the interplanetary magnetic field grew progressively weaker. The intensity of the cosmic rays, however, increased steadily with distance from the centre of the solar system.

... observations—made way out in the solar system by a satellite launched 20 years ago—have allowed us to unravel what the Sun was doing hundreds of years ago.

The experiments measured the intensities of the protons, as well as the helium, oxygen and other heavier elements in the cosmic radiation. Magnetic fields have a stronger influence on protons than on helium and the other heavier cosmic rays, and this provides an important clue to the workings of the outer solar system. The manner in which the proton and helium cosmic rays increased as the satellite receded from the Sun greatly improved our understanding of the manner in which the Sun prevents cosmic rays from reaching Earth. This in turn, allowed us to improve our interpretation of our beryllium results (see Chapter 32).

These observations—made way out in the solar system by a satellite launched 20 years ago—have allowed us to unravel what the Sun was doing hundreds of years ago. Many different scientists were involved and together their observations yielded results that none could have achieved by themselves.

The two *Pioneer* satellites were ultimately too far from Earth for their radio signals to be understood. The last data they sent back, in 2002, made it clear that they had not left the solar system.

The *Voyagers'* radio transmitters were more powerful, so we turned our attention to them. How soon, we wondered, will they cross the Termination Shock, the region of space where the solar wind comes to a screeching halt? Theory told us that the solar wind would slow down, become denser and that the magnetic field would become four times stronger. As far as the energetic particles and cosmic rays were concerned, well there were lots of arguments about what they would do. Unfortunately, the solar wind instrument had broken down on *Voyager 1* some years earlier, so some of the best evidence would not be available.

Finally, 26 years after it was launched, at a distance of 92 astronomical units (92 times the distance from Sun to Earth), *Voyager 1* sent back some very unusual results. There were many more energetic particles than before. The magnetic field had increased, but not by a factor of four. Controversy raged. Some thought *Voyager 1* had passed through the Termination Shock. Some didn't.

To confuse the matter even more, the energetic particles went back to normal after several months. Those who said that spacecraft had not gone through the Termination Shock claimed that they had been vindicated. The other camp stated that the observations simply meant that the Termination Shock had a wavy surface. No one was satisfied.

Voyager 1 silenced the argument six months later. Lots of energetic particles were seen again. The magnetic field obliged by increasing by a factor of four. Everyone agreed that now, at a distance of 94 astronomical units from the Sun, *Voyager 1* had passed through the Termination Shock. Two years later, in 2007, *Voyager 2* did likewise.

The *Voyager* spacecraft are not completely out of the solar system yet. They are now in the region called the heliosheath, beyond which is the Bow Shock, estimated to be in the vicinity of 150 astronomical units away. Once through that they really will be in interstellar space. That will take another twenty years, making it almost 50 years after the *Voyagers* left Earth. Few of the scientists and engineers who designed and built them will still be alive.

That will take another twenty years, making it almost 50 years after the Voyagers *left Earth. Few of the scientists and engineers who designed and built them will still be alive.*

In the early days of space research, we knew little about what was out there and our satellites gave us quick answers. With the passage of time, the instruments have become much more complicated and expensive, and our observations need to continue for a decade or more to provide the information we seek. Few of the scientists analysing the data on modern satellites have seen the instruments they are using—their only contact is via a stream of numbers. It is a very different world than the hands-on, risk prone days at the beginning of the Space Age.

‹ 34 ›

Feral physicist farm

When faced with a knotty problem, I tell my colleagues that I am going out to 'have a chat about it with my bull'. Despite the inevitable jokes that they are not paying me for bullshit, they know that it works.

In the late 1980s, Gillian and I began to do a little bit of long-term planning. Where, we thought, will we retire in ten years time? Will it be Tasmania, South Australia, New South Wales or somewhere else?

At that time I was very stressed in my job in CSIRO and Gillian decided that a short break in the country would be a good idea. She booked us in for a weekend at a palatial hotel near Bowral, in the southern highlands district of New South Wales.

The date is firmly fixed in my mind; it was the 30th anniversary of the flight of *Sputnik 1* on 4 October 1957. As a consequence, our relaxing time was interrupted by my preparation for and participation in a long interview on the ABC in recognition of that event.

That over, we started to drive around the Southern Highlands and were immediately captivated by the place. Its rolling hills and bracing climate were reminiscent of our childhood in Tasmania and our years in Boston. It had a gaggle of small villages, many dating back to the earliest days of European settlement. Finally, it was halfway between Canberra and Sydney. I was working in Canberra at the time and Gillian in Sydney, so it didn't require any genius to work out that it would be a good place to meet at the weekends.

True to my background, we prepared a 'specification' of the property we wanted to buy. For the garden and farming, it needed volcanic soil, good water and a north-easterly facing aspect, as well as mature trees, an oldish yet substantial house, good views and so on, to a list of ten different attributes.

The Sydney real estate market was experiencing a boom at the time, real estate agents were being run off their feet and properties were selling within days of coming onto the market.

The real estate agents in Bowral were less than enthusiastic when shown our 'specification'. More so when I would take a geological map, a stereoscope and a pair of aerial photographs out of my bag to 'inspect' the full extent of the properties we were being shown. Frequently I could see that the real estate agents were being a little economical with the truth. They didn't seem to appreciate being told so.

At first we didn't find what we wanted and settled for a rundown, 100-acre farm with a rather horrible little house and a hay shed that had been turned into a revolting granny flat. We renovated the house over a frantic week, rented it out and started to use the granny flat for ourselves.

We bought ten steers to begin practising at being farmers. Unfortunately, in our ignorance, we bought one completely mad and uncontrollable steer, which we called Toro Loco.

Six months later we were shown a property that exactly matched our specifications, except it was 850 acres in area, roughly four times what we had in mind. Nevertheless we bought Jellore, and sold our earlier property at a 50 per cent profit. Risks sometimes work out very well.

Having bought it, we began to learn more about Jellore. It is named after Mount Jellore, the second highest mountain in the Southern Highlands and our property goes to within 30 metres of the top of the mountain. It is an old volcanic plug and sticks out prominently from the surrounding hills. It can be seen from a great distance. For this reason, It was used by Major Mitchell, the second Surveyor-General of the colony, in the preparation of his 1828 'trigonometrical survey of the nineteen counties of NSW'.

For several years following its purchase, I held several small remote sensing conferences at Jellore. Surveying was high technology for 1828 and an early forerunner of remote sensing, so I pointed out to the participants that they were meeting at place of real historical significance. 'It was', I said, 'the birthplace of remote sensing in Australia'. As usual, I received a few queer looks and suggestions that 'I pull the other one'.

I was also lucky enough to be presented with one of the earliest *Landsat* images, a photograph of Jellore Farm from a satellite at an altitude of 800 kilometres. It and several instruments, from *Pioneer 6* and my *Interplanetary Monitoring Platform* experiment, occupy prominent positions in the farmhouse.

In the eighteen years since we moved to Jellore, I've returned to space science. As described in an earlier chapter, I advised on the development of a remote sensing in the mining industry and returned to studies of the Sun and interplanetary space.

A farm is a great place for intellectual activity; it provides the means to overcome mental blocks. When faced with a knotty problem, I tell my colleagues that I am going out to 'have a chat about it with my bull'.

Despite the inevitable jokes that they are not paying me for bullshit, they know that it works. While walking the cattle into a new paddock or forking out hay during winter a new idea or way around a problem will suddenly flash into my mind.

As far as I can tell, not many space scientists own and manage farms in Australia. This sometimes leads to interesting situations. For example, during the height of the recent drought of 2002–06, several cows died suddenly. My local vet and I wondered if they had been poisoned by the algae growing on all our dams at the time. He decided to call for the big guns, the veterinary specialist from the Department of Agriculture.

As we inspected the dead cows and poked around in their various bodily orifices, we chatted amicably about the drought and this and that. After a while, the vet said, 'You talk like a scientist. Are you a graduate in agricultural science?'

'No', said I, 'I'm a rocket scientist'.

He stared and said, 'You're bullshitting me?'

Having set his mind at rest in that regard, we chatted on at length about global positioning systems, remote sensing and the scientific reasons for climate change.

Sadly, the evidence for climate change is clear to see on our farm. In the 1920s vegetables grew on a paddock that struggles to grow grass. The previous owner dug an irrigation trench—which is still visible—from a creek that now flows at an impossibly low rate for vegetables or any other intensive crop.

Dicksonia antarctica (tree ferns), for example, are growing under conditions of great stress in a number of locations on the farm and, each summer, one or more die. The climate they knew at birth is long gone.

In the past Gillian and I have found that her work—in the arts—and mine in science sometimes reinforce each other in a remarkable manner.

Here's an example. In preparation for a future exhibition, one of her associates showed us historical records of the depth of water in Lake George, near Canberra. From these we learned that the lake was 12 metres (40 feet) deep when Governor Lachlan Macquarie visited it in 1821. By 1850 it was as completely dry as it was during the drought of 2002–06. By 1880 it was almost as deep as it was in 1821. It then dried up completely again before 1900.

All of this was happening long before the 'greenhouse effect' of global industrial activity increased carbon dioxide levels in the atmosphere. Interestingly, these changes in Lake George coincide quite nicely with the changes in the spottiness of the Sun in the 19th century …

Scientific research is an addictive activity to some people. The riddles posed by nature attract us and challenge us to document, analyse and understand them. Retirement is a concept I really do not welcome or understand.

Living on our farm, Gillian's and my professional activities combine and challenge us to understand the changes in the climate of Earth over the past millennia, centuries and decades. We like to think that our insights might, just might, help to throw light on this very complex question.

‹ 35 ›

Our fickle Sun

When the Sun was not spotty, cosmic ray intensity was high and the Earth's climate was colder than usual. When the Sun was spotty, cosmic ray intensity was low and the Earth was warm. It happens too often to be a chance correlation of two independent phenomena.

The Space Age has taught us a great deal about the Sun. In particular, it has shown us that it is a restless, moody star, which has exposed the Earth to influences that we could not understand before. Our studies have shown us that it may have some nasty shocks in store for us in the future.

In completing my story, I briefly outline what we have learned about our nearest celestial neighbour and how it may influence us here on Earth in the future. No scientist works in isolation; progress is usually made from the synthesis of many measurements and the ideas of many people. This is therefore a summary of what we have learned collectively over the past 50 years.

We now know that sunspots, which Chinese and Korean astronomers were observing over 1000 years ago, are regions of the Sun's surface where there are strong magnetic fields—so strong that the Sun's surface is cooler and less bright.

The number of sunspots waxes and wanes every eleven years, and the peak values of this oscillation vary from decade to decade. For example, few sunspots were seen for the 50 years between 1650 and 1700, while very large numbers were visible in 1958, 1970, 1981 and 1991. Why the Sun's magnetic fields change over time like this we don't know for certain, yet.

As described in Chapter 14, the MIT experiment in 1960 showed that there was a fast-flowing solar wind that blows away from the Sun. That wind has quite unusual properties, including that it drags magnetic field along with it, sometimes way out past the orbit of Earth. As the spottiness of the Sun varies from year to year, so the strength of the magnetic fields throughout the solar system varies with time. The stronger the magnetic fields in the solar system, the more cosmic rays are prevented from reaching Earth—so more sunspots, more interplanetary magnetic fields, fewer cosmic rays at Earth.

Following the pioneering observations of Carrington and Hodgson, we now know that solar flares occur frequently in the vicinity of sunspots. They often eject enormous chunks of solar material away from the Sun at great speed, sometimes as high as 2500 kilometres (1550 miles) per second. We call them coronal mass ejections (CME) and they have profound effects upon the space between the planets. When they slam into the Earth's magnetic field, they produce a 'magnetic storm'. Compasses swing back and forth; pigeons are said to be unable to find their way home. Strong displays of the aurora are seen in the polar regions. In a really large storm the aurora can be seen a long way from the poles; aurora were seen across all of southern Australia during the big magnetic storms of the International Geophysical Year in 1957–58.

Magnetic storms sometimes have a major impact upon our technological society. On several occasions in North America they have induced large electrical currents into electrical transmission lines, which are the lifeblood of modern society. On these occasions the fuses on the transmission lines have blown in 25 per cent of the country and it has taken up to a day to get the electricity flowing again. This has resulted in major commercial losses. Electrical currents are also induced into gas pipelines, hastening the corrosion that will ultimately cause them to leak.

As solar flares and coronal mass ejections accelerate cosmic rays, sometimes the intensity of this cosmic radiation is so great that it causes permanent damage to the fragile solar cells that power most satellites and keep their electronic systems working. This is particularly serious for communications satellites. A big burst can reduce the lifetime of a satellite by one year. That's a lot of lost profit for the operators.

One might hope that all the satellite measurements we've taken would have shown us all the tricks the Sun can play on us. The ice cores from the polar regions tell us otherwise. It's clear that the Sun has been extremely active for the last 50 years; the evidence indicates that it has been one of the most active periods in the last 10,000 years. For perhaps 20 per cent of the last 10,000 years, there were no sunspots at all. This suggests that sooner or later the Sun will become less active. Some scientists think that this could happen in the next ten years or so.

Logic might say this would be good. Less solar activity should mean fewer magnetic storms to shut down electricity systems and corrode pipelines, and fewer bursts of cosmic rays from solar flares to damage communications satellites or threaten astronauts' health.

On the other hand, we know that more cosmic rays reach Earth—from out in the galaxy—when the Sun is less spotty. Which of these conflicting effects will win?

Personally, I think the evidence suggests that the Sun will soon go back to the activity levels of about a century ago, with more magnetic storms and big radiation bursts, meaning more problems for technology here on Earth.

The Sun, however, isn't always logical. The historical evidence indicates that magnetic storms and big bursts of radiation were more frequent about a century ago, when the Sun was at an intermediate level of activity. There are theories to explain this, but we really don't know the cause for certain.

Personally, I think the evidence suggests that the Sun will soon go back to the activity levels of about a century ago, with more magnetic storms and big radiation bursts, meaning more problems for technology here on Earth.

These effects are now referred to as 'solar weather'. As with the weather on Earth, we have learned to predict bad solar weather for up to ten or more days in advance. The position and size of the sunspots are useful predictors of the solar weather. These are what we used during the *Apollo* flights to the Moon.

We now have instruments on satellites that can see the X- and gamma rays from solar flares and coronal mass ejections as they leave the Sun. Together, they allow us to predict how much and how soon the solar weather will affect the Earth.

There are always more questions. Looking at the cosmic ray record over the past 1000 years, there is an uncanny similarity to the changes in the Earth's climate. When the Sun was not spotty, cosmic ray intensity was high and the Earth's climate was colder than usual. When the Sun was spotty, cosmic ray intensity was low and the Earth was warm. It happens too often to be a chance correlation of two independent phenomena.

There are theories of course, such as: when the Sun has more spots it emits higher heat radiation, which makes the Earth a little hotter or that when the Sun is less spotty the higher intensity of cosmic radiation produces more ionisation in the atmosphere which in turn makes the atmosphere more cloudy and the Earth cooler. In addition to all this, there's the steadily increasing 'greenhouse effect', which will certainly help warm our planet.

The big question is: 'How big are each of these effects and how much bigger will they become?'

The arguments have become very acrimonious between the 'greenhouse supporters' and their opponents, the 'greenhouse sceptics'. Each side claims that their arguments are scientifically sound and proven. It seems to me that the matter is more complex than either side would have us believe.

The level of sunspot activity should decrease soon. When it does it will help us to resolve the argument. Ultimately, nature always has the last word.

‹ 36 ›

A few reflections

Space science has changed greatly compared to when I built my satellite instruments in the 1960s. Then a satellite would be launched within two or three years from the start. Many experiments were built by a professor or two and university students working in old university buildings and the occasional army hut.

In a reflective moment, I look back over the past 50 years and contemplate the remarkable luck I have had. 'How was it,' I ponder, 'that a rather timid student from Tasmania ended up launching experiments on seven interplanetary spacecraft before he was 35?' But also, 'Why are there so many new things left for me to discover?'

When thinking like this, I realise the enormous advantages gained by being a pioneer in any new area of science. Back in 1957 mankind knew very little about space. With almost anything we did we discovered something new.

There was a downside to this. The rockets were unreliable and many years of work might be wasted because of a fault. The career of a university scientist, then as now, depended on their ability to publish 'important' scientific papers about their research.

You couldn't publish good science if your rocket crashed. It made space science a risky area to go into.

Furthermore, many scientists found the complete lack of information about space quite threatening. They feared ridicule, or high levels of criticism, if they proposed experiments to be flown in space. As a consequence, few were prepared to enter this fledgling field.

This, strangely enough, was an important piece of 'good luck' enjoyed by four young Australians, including myself, who became space scientists in America soon after the flight of *Sputnik*. We didn't have reputations and we didn't mind risk. Most of us had worked in the Australian Antarctic Program and knew how to make scientific instruments operate under adverse conditions.

Deep down, we were betting the chances that we would make an important discovery were greater than not. We also were acting in the certain knowledge that 'bugs' would eventually be ironed out of the rockets, and that they and the satellites would become bigger and much, much smarter. History shows that we were right.

Another part of our 'luck' was that there were enormous advances in technology in the 1950s and 1960s. We happily employed these new 'tools' to do things that were quite impossible in the late 1940s when I left high school. The computer and solid-state electronics (the transistor, microprocessor and later the computer chip) are the more obvious of these. There were many more; new materials that emit a flash of light when a cosmic ray passes through them and the photomultiplier, a kind of light amplifier, came along at just the right time for me.

The early years of space exploration in America relied greatly on 'hands-on' science, which gave us Australians a great advantage.

Our backgrounds in the Antarctic program and elsewhere had taught us how to design, build and trouble-shoot electronics. We had all learnt to improvise—the hacksaw, soldering iron and sticky tape were our tools of trade. We were accustomed to making split-second decisions without going away to look up information in textbooks or asking for other opinions. This was all second nature to us—we could concentrate on the more important scientific issues while many of our peers were still struggling with their uncertainties about what to do.

Another part of my 'good luck' was the Australian education system. It was rigorous, and gave me a solid background in mathematics and the scientific method. Once I overcame my initial nervousness on arriving at the Massachusetts Institute of Technology—and meeting respected people like Bruno Rossi and Jim van Allen—I realised there was little that I could not do.

Then, too, there were teachers, mentors and advocates whose contribution was invaluable. There were my mathematics and science teachers at high school who encouraged me and tolerated my adventurous nature—particularly in chemistry. There were my colleagues at the University of Tasmania and, particularly, my supervisor Geoff Fenton, who obtained all the money needed for me to construct and establish my equipment in Papua New Guinea, Tasmania and Antarctica.

Finally there were the scientific leaders like Bruno Rossi and Frank McDonald who helped me, explained things that I did not understand and who made certain that their friends knew about me. This was good luck, indeed.

Space science has changed immensely compared to when I built my satellite instruments in the 1960s. Then a satellite would be launched within two or three years from the start. Many experiments were built by a professor or two and university students working in old university buildings and the occasional army hut. Everything was done in a hell of a hurry. We built several spare instruments in case something went wrong.

Once the satellites were successfully launched, the data from the experiments was 'owned' by the scientists who had designed and built the instrument. They had total control over who was allowed to see or use the data.

It is totally different now. The instruments are usually built by large teams of scientists from many different laboratories and often in many different countries. Many will never see the actual instrument that is flown into space. The instrument is usually designed and constructed by a specialist space-engineering company. Sometimes, the satellite will not be flown until twenty years after the start of the program. Frequently, the data is available on the internet within weeks for all to use.

At the dawn of the Space Age, most satellites were designed to investigate a number of different areas. *Pioneer 6* was typical of the satellites of that era. Now we have specialist 'space observatories', which operate just like ground-based astronomical telescopes. The *Hubble* space telescope is the most famous of these, but there are many looking at the galaxy and the Sun in X-rays, ultraviolet, infra-red and radio waves.

These are large, very expensive satellites that are controlled by a team of technicians and engineers situated in a central satellite-control station. Astronomers are allocated 'observing time', a set period during which they point the space telescope at whatever star they are studying. They do it all from the computers in their offices, sometimes on the other side of the world. The data is sent back from space by radio, deciphered and sent on to the astronomer's computer so that he or she can check the results in 'real time'.

The modern world of space is very professional and efficient. Certainly, young scientists still make marvellously new discoveries, which challenge our accepted knowledge and update our textbooks. Certainly, they do not have to wait for the satellite to be launched; they usually go onto the internet and access the data they need from one of many international data archives.

This type of research, however, has little of the feeling of personal involvement, adventure and imminent disaster that we had at the dawn of the Space Age.

I greatly treasure my good luck that I happened to be in the right place, at the right time, to participate in the dawn of the Space Age to the full.

‹ 37 ›

The next 50 years in space

Let me now peer into my rather cloudy crystal ball and speculate on what advances in space technology and exploration may occur over the next 50 years.

Many of those changes will be driven by commercial and military requirements. Companies and governments will develop bigger and better communication, remote sensing and navigation satellites. They will be needed to provide 'real-time inputs' for the 'information age'. They will provide the successors to GoogleEarth—which allowed a friend in Switzerland to count the cows in my paddocks with a click of his computer keyboard—or the pedantic Global Positioning System (GPS) receivers that speak to us in our chosen language or accent, to tell us with mathematical precision which street to turn at en route to our desired destination.

Soon, these technologies will go much further. Soon they will allow our passenger aircraft to take off and land automatically, without input from a pilot. Already there is a joke in the aviation industry about the commercial aircrew of the future being a pilot and a Rotweiller.

The aircraft flies itself using data from GPS and communications satellites. The pilot's job is to feed the Rotweiller. The dog's job is to bite the pilot if he touches the aircraft controls.

Don't laugh. The mining industry has been working for several years to automate the hundred or so massive 300-ton trucks that run around 'open cut' mines, and dispense with the drivers. American railway companies are investigating automating their long distance trains. Undoubtedly, new and smarter satellites will result in fewer high-paid, hands-on jobs on Earth.

For the past decade experimental 'sentinel satellites' have been parked a million or so kilometres away from the Earth, to give warning of the coronal mass ejections that cause the worst space weather.

As I discussed in chapter 35, the Sun generates 'solar weather' that can play merry hell with our modern technologies. Irreversible damage to communication satellites, shutting down large power grids and interfering with military radars are all of great concern, worldwide. Of course, our mining companies wouldn't want all their trucks to run off the road and tumble down into the mine.

For the past decade experimental 'sentinel satellites' have been parked a million or so kilometres away from the Earth, to give warning of the coronal mass ejections (CME) that cause the worst space weather. We will see greatly improved sentinel satellites over the next few decades. They will allow us to predict the space weather and greatly reduce its commercial and other impacts on our technology.

The forerunners of these new watchdog satellites were launched in 2005. Two 'stereo spacecraft' were launched into orbits very similar to that of Earth—one travelling around the Sun behind Earth, the other ahead. They carry coronagraphs—instruments that create an artificial solar eclipse, allowing us to observe the outer atmosphere of the Sun (the corona). They therefore provide two views (a 'stereoscopic pair') of a CME as it leaves the Sun, providing a much greater ability to predict the consequences of the CME when it hits Earth two days or so later.

Other sentinel spacecraft will be parked at the 'Lagrangian Points'; where the gravitational effects of the Earth, Sun and Moon balance and the sentinels will stay fixed in space relative to Earth. They, together with future stereo spacecraft, will allow us to anticipate and minimise the effects of our moody neighbour, the Sun.

There will be other advances that are not driven by commerce or the need to protect our technological infrastructure. To discuss them, let us first consider the reasons why mankind has in the past—and will in the future—spend enormous amounts on space research and space travel.

From the beginning, space research was used as a test-bed for advanced technologies used by the military and society, and as a demonstration of national might. For example, the very precise navigation systems that guided *Pioneer* spacecraft provided the experience and expertise to improve the aim of military rockets.

This learning process continues. Space is now the test-bed for automatic control systems and artificial intelligence like those that allowed the mobile laboratories *Spirit* and *Opportunity* to travel safely on the surface of Mars. More and more countries are entering into this test-bed business in order to gain access to the technologies that the pioneer space nations now refuse to sell them.

This need to develop technologies, and nationalistic ambitions, will lead more nations to embark on manned and unmanned space programs. China has already launched its own astronauts into orbit on its own rockets and is planning for the time when it will send astronauts to the Moon. India will fly its first 'recovery capsule' in 2008, which will allow them to launch an Indian astronaut in the relatively near future. Recently, there was a debate on Indian television about their plans to go to the Moon.

The Americans have directed a great deal of study towards establishing a semi-permanent outpost on the Moon as well as sending astronauts on a 18-month mission to Mars and back.

Of these two, the lunar outpost seems most likely in the next two decades or so. NASA has made very detailed studies of how it would be done. It would probably be located near the north or south pole of the Moon, where the evidence suggests there may be some water, and where the temperature extremes will not be as great as at lower latitudes.

The samples of lunar rock obtained by the *Apollo* missions in the 1970s showed that they were extremely rich in a rare form of the helium atom, ^{3}He, with two protons and only one neutron in the nucleus (helium usually has two neutrons and two protons in the nucleus).

For the past 50 years, without conspicuous success, nuclear physicists have been trying to make energy in a controlled manner using 'nuclear fusion'—the process that occurs in the Sun and in the hydrogen bomb. The rare form of helium, ^{3}He, is their favourite for the fuel that would be used to generate electricity from controlled nuclear fusion—if it is ever made to work.

As I have emphasised in earlier chapters, space is a risky business. So, undeterred by our inability to make nuclear fusion operate in a meaningful manner, one of the goals of an American lunar outpost will be to mine ^{3}He from the lunar rocks and bring it back to Earth to provide energy from nuclear fusion should we decide to stop burning fossil fuels in the future.

Clearly this idea is really 'out of the box'—but probably no more so than the idea of space flight in the years prior to the flight of Sputnik.

Several years ago, NASA called for proposals from industry to conduct design and feasibility studies for a lunar outpost and lunar mining at industry's own cost. Perhaps, I thought, they might have a hundred or so proposals submitted. Wrong—there were more than 6000. Some were from international mining companies. Clearly there is a great deal of interest from both governments and industry, and it may well happen.

Mars, however, is the destination that excites the greatest interest on the part of space enthusiasts. It has a weak atmosphere; it seems to have frost and water at its poles; there is clear sign of erosion by water in the past; it has a magnetic field and the solar illumination is 45 per cent of that at Earth, so solar cells will work. An occupied Martian research station would be challenging, but possible.

Scientists see Mars as a sister planet of Earth and capable of telling a great deal about the evolution of planet Earth. The technology development people like it too: radio signals can take up to 20 minutes to go from Earth to Mars and they may be quite weak on arrival. This will makes 'real-time' control, and 'wide-bandwidth' television impossible.

Consequently, missions to Mars offer a wonderful reason to develop automatic reconnaissance vehicles, conduct research into artificial intelligence and very high bandwidth communications based on lasers.

Only a well funded space program could afford to develop the things the technologists dream about.

For these and other reasons, the exploration of Mars will be one of the *really big* space activities over the next 50 years. Unmanned mobile laboratories will do most of this. The American and European robot exploration vehicles will be joined by Japanese, Chinese, Indian and Brazilian vehicles and those of other upwardly mobile nations. There might even be traffic jams at the points of greatest scientific interest.

The human race will always speculate about going there. NASA has made various studies—not least being how to reduce the risk posed by the galactic and solar cosmic radiation that started my whole career in space (Chapters 7, 14, 24). They talk about using the water supply on a spaceship as a 'storm-cellar'—a small room surrounded by water to shield astronauts from the radiation. The use of pharmaceuticals that accelerate the body's ability to repair radiation damage is also being considered.

Recognising the heightened risk to the human reproduction system (Chapter 24), it is possible that only more mature astronauts would be considered to go on long space voyages. In the early days of the Australian Antarctic program it was usual to surgically remove the appendix before going south, to eliminate an important medical risk.

I wonder if they have something like that in mind for the Martian astronauts?

Closer to home, there will be an inevitable increase in 'space tourism'. A number of tourists have already flown to the International Space Station on Russian rockets. There are several companies developing rockets that will be launched from high-flying aircraft. They will be used at first to provide space tourists with sub-orbital flights of 15 minutes, which will undoubtedly lead to longer flights of several orbits of the Earth over the succeeding decades.

Space tourism is, of course, an exaggerated form of boasting. The space tourism companies will offer more and more ambitious tours to attract those people who have a lazy ten million or so to spend.

Who knows, sooner or later, there may be a 'Restaurant at the End of the Universe'—perhaps at one of the Lagrangian points—taking its name from that renowned establishment in Douglas Adam's book: *The Hitch Hikers' Guide to the Galaxy.*

Sources

Cover (both images NASA).

Plate 4a. (NASA)

Plate 4b (Royal Swedish Academy of Sciences)

Plate 8 (DSTO).

Other images Ken McCracken collection

Figure 1 from an article by P Meyer, E N Parker and J A Simpson, in *Physical Review*, 104, 768, 1956 (American Physical Society).

Figure 2 thanks to M A Shea and D F Smart.

Figure 3 from an article by K G McCracken, *Journal of Geophysical Research,* 67, 423, 1962. (American Geophysical Union)

Figure 4 from an article by R P Bukata, U R Rao, K G McCracken, and E P Keath, *Solar Physics,* 26, 223, 1972 (Springer Science and Media).

Figure 5 from an article by W C Bartley, R P Bukata, K G McCracken and U R Rao, *Journal of Geophysical Research*, 71, 3297, 1966.(American Geophysical Union).

Figure 6 from an article by R P Bukata, U R Rao, K G McCracken, and E P Keath, *Solar Physics*, 26, 223, 1972 (Springer Science and Media).

Figure 7 from an article by K G McCracken and N F Ness, *Journal of Geophysical Research*, 71, 3315, 1966. (American Geophysical Union)

Further reading

Heppenheimer T A, *Countdown: a History of Space Flight,* John Wiley and Sons, New York 1997.

Dickson P, *Sputnik: The Shock of the Century*, Walker and Co, New York, 2001.

Dougherty K and James M, *Space Australia: the story of Australia's involvement in space,* Powerhouse Publications, Sydney 1993.

Cadbury D, *Space Race: the Epic Battle Between America and the Soviet Union for Dominion of Space*, Harper Collins, New York 2006.

Burnett T, *Who really won the space race*, Collins and Brown, London, 2005,

Carlowicz M J and Lopez, R E, *Storms from the Sun: the emerging science of space weather,* Joseph Henry Press, Washington USA, 2002.

Soon W Wei-Hock and Yaskell, S H, *The Maunder Minimum and the variable Sun-Earth connection*, World Scientific, Singapore, 2003

Eddy JA, *The case of the missing sunspots*, Scientific American, Volume 236, May, 1977

Parker E N, *The Sun*, Scientific American, Volume 233, 42, 1975.

About the author

Ken McCracken commenced his career as a fledgling space scientist at the University of Tasmania in 1954. He made important contributions to the International Geophysical Year in 1957-8 and, as part of that process, established a cosmic-ray laboratory in Papua New Guinea in 1957.

Two years later he was invited to join one of the world-leading space research laboratories at the Massachusetts Institute of Technology (MIT) in the United States. Over the following six years he designed and built instruments for the interplanetary *Pioneer* spacecraft that NASA flew to the orbits of Venus and Mars. This allowed him to make important discoveries about the nature of the radiation and magnetic fields in the solar system.

McCracken was closely involved in protecting the American astronauts from possibly fatal doses of radiation during their exploration voyages to the Moon.

Returning to a professorship in Physics at the University of Adelaide in 1966, he and his colleagues discovered bright X-ray stars in the southern sky using *Skylark* rockets flown from the Woomera rocket range in South Australia.

In 1970 CSIRO appointed McCracken to establish a new laboratory in Sydney. His brief was to develop improved methods to discover deeply-buried ore bodies. Fourteen years later CSIRO asked him to develop the CSIRO Office of Space Science and Applications and the Australian government appointed him an inaugural member of the Australian Space Board.

In retirement, McCracken is a visiting senior research fellow in space research at the University of Maryland and a frequent visitor to the International Institute of Space Science in Bern, Switzerland. He is an Officer of the Order of Australia, a co-recipient of the Australia Prize and a fellow of the International Academy of Astronautics, the Australian Academy of Science and the Australian Academy of Technological Sciences and Engineering.

He and his wife operate a cattle-breeding farm in the Southern Highlands of NSW, where he can indulge in his recreational passions that include bush-walking and amateur radio.